Small-Scale
PIG RAISING

"A pig in almost every cottage sty! That is the infallible mark of a happy people."

William Cobbett, *Rural Rides*

"There are very few mixed farms in the country on which a few pigs of one type or another could not be kept to advantage. The presence of a few pigs does not increase the labour bill, and they do not reduce the capacity of the farm for carrying other forms of stock. By-products such as chat (small potatoes) tail corn and surplus green crops are invariably available, thus reducing food costs. The result is that a small pig unit makes a profit which is out of all proportion to its size."

V. C. Fishwick, *Pigs: Their Breeding, Feeding and Management*

Small-Scale
PIG RAISING

Dirk van Loon

STOREY BOOKS
Schoolhouse Road
Pownal, Vermont 05261

The mission of Storey Communications is to serve our customers
by publishing practical information that encourages personal independence
in harmony with the environment.

"The Order of the Golden Paste Pot" goes to the editor, Walter Hard, Jr., who also snipped and pasted his way through the disorganized masses of paper that became *The Family Cow.*

Small return, Walter, for your efforts on my behalf these past years; for your support, interventions, and above all for your infinite patience. My thanks to you.

Copyright © 1978 by Dirk Van Loon.

The information in this book is true and complete to the best of our knowledge. All recommendations are made without guarantee on the part of the author or Storey Books. The author and publisher disclaim any liability in connection with the use of this information. For additional information, please contact Storey Books, Schoolhouse Road, Pownal, Vermont 05261.

Storey Books are available for special premium and promotional uses and for customized editions. For further information, please call Storey's Custom Publishing Department at 1-800-793-9396.

Printed in the United States by Capital City Press
40 39 38 37 36 35 34 33 32

Library of Congress Cataloging-in-Publication Data

Van Loon, Dirk.
   Small-scale pig raising.

   Bibliography: p.
   Includes index.
   1. Swine. I. Title.
SF395.V26   636.4'08   78-12938
ISBN 0-88266-136-1

Contents

Introduction   vii

1 The General Pig   1

2 The Wild Boar   7

3 History of Domestication   13

4 Behavior and Form   30

5 Buying a Piglet   45

6 Handling Pigs   53

7 Penning a Pig   63

8 Furniture and Utensils   79

9 The Meat Producers:
Digestion and Physiology   84

10 Nutrition   92

11 Feeds and Feeding   102

12 The Scavenger Pig   114

13 Rearing Your Own Piglets   131

14 Management Techniques   150

15 Health   169

16 Butchering   194

17 Portioning, Processing,
    and Curing Pork   220

    Biliography   240

    Glossary   242

    Appendix   244

    Index   259

Introduction

In writing this book I have tried to create the kind of how-to *and why* guide I wish I'd had when we first raised pigs.

First of all, it is a book for beginners. However, I have gone further and deeper into the subject than introductory books normally do. There is a lot of history here, as well as glimpses at swine management practices through the centuries.

To write this book I have relied heavily on the experiences and writings of others who have devoted much more of their lives to raising swine. Therefore, if *Small-Scale Pig Raising* is successful, we owe a great deal to the many writers whose books and extension publications I have read, to the numerous people in universities and extension services across North America and in England who have answered my letters, and to those farmers and teachers closer to home who have given me hours of their time answering endless questions. I am grateful to all of you for your help.

To John and Gloria Corbett, farmers in Shelburne County, Nova Scotia, and Angus Rouse, swine technician with the Nova Scotia Department of Agriculture and Marketing, special thanks for repeated assistance over the past two years that this book has been developing. Also I am once again indebted to Craig Wheaton-Smith of Dorset, Vermont, for his generosity in loaning me books from his fine agricultural library, and to Drs. Jean and Gene Ceglowski of West Rupert, Vermont, for their comments and criticisms on material dealing with swine health.

DIRK VAN LOON

CHAPTER 1

The General Pig

The other day a friend considering raising an animal for meat asked whether I thought he would be better off getting a calf or a piglet. Without hesitation I said pig. A pig makes more sense for a person who has never raised an animal before, for one who has not got much land, and for one who wants to accomplish the most in meat production with a minimum investment of money and time.

Why Raise a Pig?

LOW RISK/SHORT TERM

Although the weaner piglet may cost twice as much as a newborn calf, it will represent far less risk. The weaner, at six to eight weeks of age, is like a five-year-old child, well started in life and able to cope without special foods or mothering. The "bob" calf is still an infant, needing milk or milk replacers and lots of care to avoid diseases that can ravage newborn animals.

Raising a piglet for pork is a short-term commitment. You're in and out of it in five or six months, free to look back and assess what happened. A flock of laying hens ties you up for a year or more. A calf for beef represents an involvement spanning at least ten months. A dairy cow is a huge commitment. So is a goat. Sheep, too.

How convenient then, with pigs, to be able to step into the business in the spring, when you're full of that special fever, and be out by Thanksgiving. If it suited you, the sty lies waiting for more and better years. If not, you can turn the building into a dog house.

Piglets for pork demand little time, day-to-day. Fifteen minutes should be average for the feeding and cleaning of two or three animals. Your own slaughtering, curing and storing may take twelve to fifteen hours altogether.

SMALL CASH INVESTMENT

The piglet-for-pork enterprise represents, overall, a relatively small investment. Perhaps $85 per pig (including housing) to start—unless you have to build from scratch with new materials—and about $150 overall. A beef calf or a small flock of chickens may cost about the same, but cows, milk goats or sheep cost much more.

Profit. These past few years it has been possible to "make" $30 to $50 in pork value over the dollar costs of raising a piglet in confinement—that is with all of the feeds being purchased, with the carcass being given the value it would have were it sold whole or by the side directly to consumers, and with labor thrown in free. No wonder some people grumble that "there's no money in it." However, there is much more that should be said about how and what is to be made with a pig.

Lower feed costs. For one thing, not all feeds need to be purchased grains. Household garbage, garden leftovers and dozens of processing by-products can be used in their place. True, they don't always have as much food value per pound as purchased, complete rations, but if the cost is low enough the difference disappears.

Too, if you keep a pig longer than the usual four or five months (with slaughtering at 200 to 220 pounds) it will make better, more economical use of many cheaper foods that have a lot of plant fiber in them. Foods like lawn clippings, cabbage leaves and pasture forages are not much use to a piglet under 75 or 100 pounds. But as he gets older he uses them well. He may not grow as fast (and you

The Swineherd.

may be a month or two longer getting him to 250 or 275 pounds weight), but the value in pork over costs of feeding can be distinctly better than those realized by commercial pig feeders. They must follow the dictates of the mass market in rapidly turning out small, lean pigs.

Building materials. I tend to ignore the costs of building materials because we are always using old boards or sheds of little or no value. But even if a sty is built with new lumber for $75, it ought to last five years or more.

LABOR

Labor can no more be given a dollar value in a small-scale piglet-to-pork venture than it can in doing your laundry. It is all maintenance time until the point that hours are being spent doing laundry or growing pork for others.

Even if you raise two or three piglets, figuring the sale of extra pork will cover the cash costs of one for your own table, it is doubtful if work time should be considered—so little extra time will be

Expenses and Gains in Raising and Slaughtering One 220 lb. Feeder Pig

Expenses		Gains	
Housing	$ 15.00[1]	Carcass, less head	
Piglet	50.00	160 lbs. x $1.25	$200.00[3]
Feed	105.00[2]	Manure,	
Medication	10.00	2.3 tons x $11.00	25.30
	$280.00		$225.30

[1] Based on $75 for housing good for 5 years.
[2] 700 lbs. commercial Grower at $15.00/cwt.
[3] Direct local sale of pork sides.

(The supermarket retail price for 140 pounds of pork—saleable cuts from 160 pound carcass—including factory cured bacons and hams is in the neighborhood of $165.00.)

involved. Often you will find that more chore time is spent walking to and from the barn than in feeding one or a dozen pigs.

MANURE

Many overlook the value of pig manure. "Good as it stinks," a farmer once told me. Yet sad to say many commercial farmers who raise pigs without a land base have no use for the stuff.

Pig manure is worth $11.00 a ton. Imagine paying to have manure hauled to a dump, yet this often is done.

A 300-pound sow will produce about 9,000 pounds of manure a year. A feeder pig, from the time it weighs 30 pounds until it is slaughtered at about 200 pounds, will produce about 4,600 pounds of manure worth $25.30 and perfect for fertilizing 1,000 square feet of vegetable garden.

**Approximate Fertilizer (Mineral) Value
of Fresh Manure from One Feeder Pig,
30 to 200 lbs. Live Weight**

Manure	Lbs	Nitrogen	Phosphorus	Potassium
(Feces & Urine) (Free of Bedding)	4666	0.6% (27 lbs.)	0.5% (23 lbs.)	0.4% (18 lbs.)

The precise fertilizer value of manure varies depending on the pig's diet and how the manure is handled. "Animals of the same kind fed more concentrates *(commercial rations are in this category)*, excrete more of the fertility elements because the food contains more," according to *USDA Yearbook of Agriculture,* 1938 (my italics). The most will be gotten out of manure that is plowed directly into the soil. Manure from which the urine has been allowed to drain away and that is piled where the rain leaches through will be worth far less than $11.00 a ton.

Smelled from a different angle—that of a dirty sty—pig manure is no asset. However, pigs do not have to be rank. When they *do* stink it is because they have not been shoveled out or because they are being forced to live in muck up to their bellies. The pig is the only farm animal that normally establishes a "bathroom" area in its pen, a practice that makes him easy to keep clean.

PIGGY BANK

For me, a clincher in the worth of raising piglets for pork on a small scale is in what I see as a forced and yet painless way of saving money. I open "savings accounts" with initial "deposits" big enough to buy small animals. From there on, deposits in feed are made on squealing demand. They are relatively small and represent

represent money I am sure I would otherwise fritter away, leaving nothing to show.

Come late fall near Christmas, when money can be tight, I close out my piglet "savings accounts," withdrawing a hundred dollars or more worth of pork per animal. It is "found" money. If I realize as little as $10 in pork value each over the dollars deposited through the summer I am doing far better than I would had I opened a proper account down at the First Savings & Loan. As for saving money in porcelain piggy banks, I've got one showing dozens of cracks from premature busts, reflecting a weakness in my own character. I really do need a piggy in the flesh.

ON PIGLETS BECOMING HOGS

Goats, especially when young, are so cute, clever and personable it is hard to impossible to slaughter those not needed for breeding or milk. Piglets, too, are cute. Fortunately, they become hogs. Just the same, remember naturalist Ronald Rood's warning: "If you're going to eat it, don't name it."

CHAPTER 2

The Wild Boar

It helps to understand and care for a pig if you know where it came from—not the breeding barn in this case, but way back to the Wild Boar that still roams about some of the more remote forests of southern and eastern Europe.

WHAT'S IN A NAME?

Boar with a capital "B" is a collective term for Wild Boars, as the word *swine* is used when speaking generally about the domestic animals. As for scientific genus and species names, all European Wild Boars and our domestic breeds of swine—there are more than 200 around the world—are given the same, *Sus scrofa*. An early Oxford Latin Dictionary says *scrofa* means a breeding sow, and says that Marcus Varro used the word that way in his book *De Agricultura* (*c.* 50 B.C.). There may be a connection between *scrofa* and the Latin *scrobis*, meaning hole or ditch. In the past this has led to one popular notion that *scrofa* had a parallel derivation with the Latin verb *scribo*, to write. Pig becomes writer, rooting up the earth in idle script. Others point to the letter-like chiselings in the pig's skin.

A few students of pigs do not go along with this same-name classification of our swine and the European Wild Boar. They call the Boar *Sus scrofa* and the swine *Sus domesticus*. But that's their argument. The animals are so much alike and may so easily be bred back and forth that it would seem that making them different species is splitting bristles. "No one can for a moment doubt that [the European Wild Boar] is the parent stock from which the domestic breeds of swine originally spring," wrote William Youatt in his book, *The Pig* (1847).

The well-known fact that all kinds breed with the boar is in itself sufficient testimony. The period of gestation is the same, the anatomical structure is identical, the general form bears the same characters, and the habits, so far as they are not altered by domestication, remain the same.

Where individuals of the pure wild race have been caught and subjected to the same treatment as the domestic pig, their fierceness has disappeared, they have become more social and less nocturnal, lost their activity and lived more to eat. In the course of one or two generations even the form undergoes modification; the body becomes

The Wild Boar.

larger and heavier; the legs shorter and less adapted for exercise. The shape of the head and neck alters; and in character as well as in form the animal adapts itself to its new position.

As Youatt indicates, the Boar is mostly a night creature. It is a deep forest creature, and being dark brown and gray to black in color, poor-eyed and long-snouted, it is well suited in style and place. The animal comes with a double coat, one a soft, wooly underhair and the other the better-known bristles that grow five and six inches long over the male's neck and shoulder. When he's mad and frightened he can raise those bristles the way a dog raises its hackles, becoming a menacing, hulking brute.

PIG VS. BOAR—COLOR AND FORM

The domestic pig's back is smoothly rounded. The Boar in profile is a craggy mountain range with peaks at the crown of its head, its shoulders and rump.

Another characteristic distinguishing Boars from domestic swine is that Boars' tails are never curly. They are bristled top to bottom, and either hang straight down or slightly bent to one side. When they are alerted, Boars clear their nostrils with a loud blow, almost a

Piglets of the Wild Boar are striped.

whistle, that is much like the warning snort made by a White Tail deer. Then they scurry for cover, their tails straight in the air.

While domestic piglets are born patterned after their parents, Boar piglets are born a light red with creamy underbellies and bold silver stripes along their sides. These stripes, like spots on a fawn, disappear within three or four months.

There is an anatomical difference between Boars and domestic swine today that either did not exist or wasn't noticed in Youatt's time, and this is the domestic animal's longer intestines. It is a feature that should make it easier for domestic swine to digest plant foods.

The Wild Boar's diet includes tree seeds and nuts, insects, insect larvae, mice and other rodents, snakes and earthworms. They eat lots of eggs of birds or reptiles, berries, fungi and carrion. The domestic hog can and will eat all of these things and more, given the opportunity to eat the many foods that modern agriculture and transportation bring together.

Wild Boars do not grow as large as domestic animals. A male may get up to 500 pounds, but for one that is purely of the wild strain and that hasn't access to farm crops, 250 pounds is more commonly tops. They may grow old, up to 30 years old, but this is no older than a domestic hog might grow were it allowed to carry on.

SOCIAL BEHAVIOR & REPRODUCTION

Wild Boars are social creatures, with an instinct to travel in groups. The same can be said for the ancestors of all domestic farm animals with the exception of the cat.

Through most of the year Boar herds are made up of three or four sows and their young, including yearling males not fully mature or not having the weight or tusks needed to compete for breeding sows. An older, larger sow will be leader of the herd. Mature boars only join herds of sows and youngsters for the once-a-year breeding season in late December and early January. One or two young boars may tag along with a middle-aged boar (there is a lot of this tag-along, learn from-your-elders behavior in wild and domestic pigs) but the oldest boars, those that some people call the Great Boars, are strictly loners except for the brief breeding (rutting) season.

Then they come a-rutting—or *ranting*, as it's called in domestic

Most of the year the senior Wild Boar is a solitary beast.

boars—and a boar "in rut" is just that, in a rut and single-mindedly pursuing the sows in heat. You can tell him from way off as he comes gruff-gruffing along with daggers of white froth hanging from his jaws. He shakes his head, sparring with bushes, treading about, urinating, and pawing the turf high like a mad bull.

The frothing comes with champing, and with the champing the male's tusks (grown from modified canine teeth) wear and sharpen against each other. The same is true with domestic boars.

Following the breeding season, male Boars leave the herds once again. Females and young continue as before until the bred sows are about to pig. At this time they, too, go off alone, looking for a secluded spot to build a nest.

Nest-building can be a fine art, even with domestic pigs. The female Wild Boar looks for shelter beneath low-hanging limbs or the crescent of an uprooted tree. If there is a cane field, she may choose that. There are stories of sows building sugar cane stacks with tunneled entrances stuffed closed with extra canes. Other more typical and less spectacular nests are mattress-like gatherings of chewed-up twigs and grasses.

The wild sow gives birth to three or four young. She can have up to a dozen, but smaller litters are the rule. The little ones stay in the nest for two or three days. Then they are out and about with mother, following wherever she leads and learning from her how to grub and root for food. They will begin feeding on their own by the age of two weeks, but continue nursing for several months.

Domestic sows have no season and may be bred any month through the year. Apparently captivity can bring this change to wild sows, or at least has among a herd of a hundred Boars on a private reserve in Nova Scotia where mothers with striped piglets may be seen trooping about any time of the year.

BOARS AND HUMANS

Wild Boars are artful dodgers and they are fast, being able to run up to 25 miles an hour for short dashes. They are tough. The males wear shields of extremely thick, gristle-like hide over their shoulders. When two of these animals spar they lean into each other, shoulder-to-shoulder, each waiting for the other to give way, allowing for a thrust and upward swipe of tusk against belly.

Where humans are concerned, Wild Boars are shy and reclusive unless cornered. Once, while on a photographing expedition, I entered a pasture with a 400-pound male. I was told that if he came toward me not to be worried. "Just stand up and wave your arms." The boar did come trotting my way, and, when he was about 30 feet off I did as I was told. The boar did not instantly charge, as some accounts of wild boar behavior would have you expect. He froze, gave me a myopic once-over, and snorted away.

Even when a Boar is wounded its first thought is escape, say people who are not out to create absurd stories of how wild things are. In most cases it is only when the Boar is cornered, or chased to exhaustion, that it turns in incredible fury on its tormentors.

CHAPTER 3

History of Domestication

Judging from the job it is to drive pigs over open ground, it's easy to guess that domestication of these animals didn't begin among nomads but had to wait until people started settling down in more or less permanent communities. That seems to be the case, and 7,000 years ago seems to be the time, out in Asia and along the Tigres and Euphrates river valleys in what is now Iraq.

Pigs and Communities

Catching the wild pigs wasn't difficult. These more or less permanent communities soon produced just as permanent piles of garbage and manure, things that were bound to lure the Boar from his forest home. No doubt one day an unfortunate piglet got its foot stuck in a discarded calabash and ended up being taken home for a pet. (Piglets tame very easily.) Domestication had begun.

THE FREE LUNCH & ITS PROBLEMS

But the Boar, wild or tame, is an awful sucker for a free lunch, and so to the present there have been many instances where it didn't pay to keep the animals confined. They could be allowed to roam free, fenced out of the gardens and fields, and could be counted on to come for regular feeds at the village dump. How simple just to wait with spear, gun, or whatever weapon, and lay down supper when hunger struck.

If a finer and fatter pig was desired, it was always easy enough to drive a young pig into a trap and to fatten it up, ho ho!

Whatever the management, for thousands of years there has been a close association between swine and people who lived in communities. Often it was a disease-ridden association, as you would have to expect considering the way pigs will root through manure. Their ability to make use of grains and other foods passed undigested through cows' systems led years ago to a practice called "following," where hogs are intentionally confined with fed cattle.

But when pigs "follow" humans, as they often did in early times, they were bound to pick up and transmit human diseases. *Trichinosis*, that disease of infestation by tiny round worms that burrow into pig or human muscle with equal pleasure, was just one threat. If there is a wonder, it's not that Moslems and Jews instituted laws against eating pork but that the others kept on, forever taking chances that their appetites would lead them to cripplingly painful diseases or death.

The Olde English Hog—not much refinement over the Wild Boar.

Speculation as to why pigs were forbidden in various cultures goes on and on without resolution. Youatt quotes Lactantius, who said that swine were declared "an abomination in the sight of God as a lesson to the Jews to obstain from the sensual and disgusting habits to which the animal is given."

Zeuner chose another fellow, Antonius, who argued that, "as the pig is valuable to the settled farmer only, the nomads, who have always felt superior, came to despise the pig as well as the farmer who raised it."

PIGS FOR SPORT AND MEAT

Although the main reason for raising pigs since domestication began has been for the production of meat, these animals have been used in a lot of other ways. Early Egyptians used muzzled pigs to trample seeds into the ground. Muzzled pigs have been used in northern Italy and southwestern France to root for truffles, a fungus that grows on tree roots.

Pigs have been saddled for riding, harnessed for driving—paired off in teams with jackasses on the island of Minorca (nineteenth century) or as teams of hogs, as in this account from Youatt:

> ...not long since the market place of St. Albans was completely crowded in consequence of an eccentric old farmer, who resided a few miles off, having entered it in a small chaise-cart drawn by four hogs at a brisk trot, which pace they kept up a few times round the area of the market place. They were then driven to the wool-pack yard, and after being unharnessed were regaled with a trough of beans and wash.... He (the old farmer) stated that he had been six months in training them.

Wild Boars and domestic pigs gone wild have provided "sport" for hunters on foot, horseback (sometimes elephant-back in India), and with dogs, spears, crossbows and guns. Then again there have been domestic pigs trained to hunt other game. Many stories of hunting pigs came out of the New Forest area in south-central England between the eleventh and fifteenth centuries when laws forbade commoners the use of dogs in their hunting.

The most famous "sporting pig" was a sow named *Slut* of the New Forest, who was trained on a lark, near the start of the last century, by brothers Richard and Edward Toomer. "Her nose was superior to the best pointer they (the Toomers) ever posessed, and no two men in England had better," it was written in *Daniel's Rural Sports*, 1807. "She stood partridges, black game, pheasants, snipes and rabbits in the same day.... She has frequently stood a single partridge at forty yards distance, her nose in an exact line."

"Slut," the sporting pig.

However, a sad thing for pigs is that no matter what they accomplish in life, it seems that their size or the quality of their hams get the last hurrahs. At age ten Slut came afoul of the law, was accused of hunting lambs, and was slaughtered. She weighed 700 pounds.

Youatt quoted a story from a *Natural History of Selbourne*, by Gilbert White, about a large and long-lived, intelligent sow who, "when she found occasion to converse with a boar...used to open all the intervening gates, and march, by herself, up to a distant farm where one was kept, and, when her purpose was served, would return by the same means."

At the age of seventeen, "by a moderate computation, she was allowed to have been the fruitful parent of 300 pigs—a prodigious instance of fecundity in so large a quadruped," White wrote, but they could not leave it at that. The story continues how this worthy sow was penned up in 1775, was fattened, and taken for measure at the dinner table where, kudos! she proved to have "good bacon, juicy and tender," and a rind that was remarkably thin.

LARD PRODUCTION

On the fat side of things, the production of lard has had its day, even as a prime reason for raising swine. There was a time in the last and into our own century when the lard a pig could produce brought a better price at market than her pork. There were several reasons for this. For one thing, growing populations of city people during the rise of the Industrial Revolution were clamoring for fats and oils of all sorts for cooking and for soaps.

Beginning in the late 1700's, fat pigs became easier to produce in northern Europe with the introduction of the blood of faster-maturing (and therefore faster to follow growth with fat) and finer-boned Chinese pigs directly from the Orient and of Neopolitans from the Mediterranean area (representing earlier crossings of *Sus scrofa* and Oriental varieties).

Then, in 1832, French chemist Michael Chevreul discovered a way to break lard down into liquid glycerine, and solid stearin compounds that could be made into inexpensive candles. For the first time in history people of little means could afford to light their homes.

These demands for soaps, for candles and for cooking fats pushed

The Chinese Pig.

Fat hogs light the world.

the price of lard, and it continued that way until the arrival of cheaper petroleum products, vegetable oils and electricity.

Bristles. Bristles have been an important hog product, too. For years Russia exported tons of them for use in brushes. Those of the top of the neck were preferred, probably because that's where they are the longest. Bristles suffer from split ends, which is why they can at the same time be tough and yet have the light touch that's wanted for painting.

Pigskin. While pig skin has been highly valued as the leather for saddles and footballs, being tough but soft and a bit spongy in texture, pig bones never gained much fame in any regard, not even as bases for stocks or soups. Earlier cultures used sheep or cow bones for tools, but not pig bones, perhaps because they develop slowly and aren't tightly knit together at the usual age for slaughter.

PIGS & SNAKES

Pigs eat snakes, and that in itself is enough reason for many people I know to raise a few oinkers. Comstock's *Handbook of Nature Study* says "it has long been noted that the hog has done a good service on our frontiers as a killer of rattlesnakes." Whether pigs are actually immune to snake venom I don't know. The *Larousse Encyclopedia of Animal Life* cautiously states that "wild hogs appear to be immune to snake bite." Certainly rooting pigs would play hell with populations of snakes that reproduce by means of eggs laid underground, but rattlers, copperheads and cottonmouths all bear live young.

CLEARING LAND

The clearing of the land of snakes is a minor thing, but it brings me around to a last (and probably most important) use of pigs over the many years since domestication began, outside of that as producers of meat. This is their use as clearers of land—forest lands—at first by accident and later by design.

In the earliest years, even before domestication was in full swing, it is likely that unusually large populations of Wild Boars or semi-wild swine were attracted to the outskirts of settlements. Large populations of hogs will destroy forests by packing the soil with their spear-like trotters, by rooting up and killing trees' surface feeder roots, and by eating nuts, seeds and seedlings until there are no up-coming generations of new trees. As the forests died back, the people could move out with fences and axes, hoes and seeds.

"Pigs prepare the way for man," wrote Zeuner, "both in regard to pasturing—for the pig can be followed by sheep as happened in the Bronze Age in Northern Europe—and in regard to agriculture."

Zeuner became even more enthusiastic at another point in his book on the domestication of animals, saying that the pig's "indirect effect on forest soils in connection with the Neolithic conquest of the wooded parts of Europe far exceeded in significance all its other technical applications."

PIGS IN THE NEW WORLD

Being short on legs and long on obstinance, the domestic pig was seldom driven far. The animal certainly never was an aid to the spread of human populations in the way of cows or horses that were easily harnessed or packed to carry goods and fed along the way on native grasses.

I imagine, though, that pigs were important aids to the spread of people by boat. Just two pigs in a canoe landed on a new shore could become a population of food in a year or two. The Pacific Islands were liberally salted with pigs this way. Christopher Columbus brought eight pigs to the New World on his second voyage, in 1493. There were cattle aboard, too, that trip, and both species thrived and multiplied. More pigs were introduced up and down the coast of the Americas by succeeding waves of colonizers, and from their success it appears they met with little competition from the native, piglike *peccaries* inhabiting warm, wet woodlands over a range extending from northern South to southern North America.

A peccary resembles a pig only at a distance. He is a much smaller animal—up to 50 pounds—and has only three toes on his hind feet instead of *scrofa's* four. The peccary's scientific name today is *Tayassu tajuca,* (from Linnaeus), but I like an earlier choice, *Dicotyles*, meaning "double navel," chosen because of the peccary's top-side scent gland just above the base of its tail. This prominent

The peccary only looks like a pig from a distance. Its compound stomach is only one of several differences.

gland looks like an extra bellybutton, and secretes a "rank musk," says Youatt. The peccary also has a complex (multi-chambered) stomach. What it lacks, and likely what put it behind pigs, is the ability to produce more than two young at a time.

Swine Management

Swine management, as it was practiced in Europe at the time of Columbus, wasn't brought over with the new pigs. Here the animals usually ran free, to breed at will and exist as best they could. This way they quickly reverted to "razorbacks," that feral type of rangy swine still common today in many southern forests of the United States, and likely still getting recharged with the blood of domestic ecapees.

Ensminger says early colonists might have provided a shed's crawl space to a pregnant sow ready to farrow, but that would be about all, in return for a harvest of fall pork taken in the wild with guns and dogs.

SWINEHERDING

In the Old World, swineherding had become a fine art, with a history reaching well back before the birth of Christ. Early Egypt had swineherds, and even the Prodigal Son was a swineherd. In Europe it became customary for a community to employ a swineherd to drive everybody's pigs to a forest each day where they would feed on whatever nature provided. Under one German system—again taking this from Youatt—people would have individual backyard sties where pigs were kept each night and fed whatever garbage or other foods there were to be had. Each morning the village swineherd would tramp through town, blowing on a horn and cracking a horrendously long whip. He might have an urchin or two as well to help gather the pigs pouring from each house lot.

Once gathered, the squealing horde would be marched from town and into some wood for a good feed. Then, with evening, back they would come, each pig knowing where to peel from the herd at home gate.

Knocking down acorns for pigs. By Jean Colombe, 15th century.
(Reproduced from The Symbolic Pig)

In other European cultures swineherding might be only a fall season job, the swineherd moving a community's pigs to an oak-beech forest where they would camp for a month or more. The swineherd watched over his grunters, saw that they didn't stray, and, with a long pole, knocked acorns and other mast to the ground.

"Fat pigs became the rage."

As commonly owned forests and pastures dwindled away, so went the swineherd. Throughout the British Isles and elsewhere came the cottager's pig: a pig confined to a sty and perhaps to an exercise yard or a bit of pasture that was granted as part of the bargain that went with working and living on His Lordship's estate.

EARLY BREEDING

Studied attempts to improve the swineherd or cottager hog through controlled breeding apparently didn't produce any appreciable results until the late eighteenth century, when the Chinese and Neopolitan hogs were imported. Up to that time regions had their peculiar varieties more or less protected by isolation, but a general lack of understanding of genetics made the development of improved pigs a slow process at best.

The Berkshire "breed," named after the county in England where this hog was found, was one of the first to reflect the impact of the Orientals. The new pigs grew, matured, and fattened more quickly.

The pig inherits this ability to lay on an insulating winter store of fat from its wild ancestors. But with the mixing of European and Oriental domestic swine, this could be taken to an extreme. Fat pigs became the rage. Other Oriental influences of shorter snouts, broader heads and "dished" faces accentuated the bulldog look.

To be large was to be admirable, and typical was an account of a

Berkshire type from Petworth, England, that measured seven feet, seven inches from the end of its snout to the tip of its tail; seven feet, ten inches around the middle; five feet around the neck and a level back that was two feet across. This was a fantastic creation so long as the lard market held. But even then a farmer sometimes had to move fast to get ahead of rats bent on taking a lard lunch from the backside of a hog so crippled with fat it could not move.

Once some of the principles of animal breeding and genetics were worked out, it took only a short time to begin creating new and improved "pure" breeds. It was an easier task—or at least faster—to improve pigs than cattle, because a sow can show her genetic mettle with a litter of eight or ten young before a cow has reached breeding age. It is easier still to breed for improved swine than for a milk cow or goat because the goal—an economically produced carcass—can be measured in all of the offspring, not just the females following their own maturing, and scrutiny through a lactation period.

On the other side of that coin, it takes no time at all for improved swine to revert to indifferent beasts if their breeding is not guarded.

BREEDING GOALS

Guarding against reversion is a job for all breeders of swine, but the guards *On High* are the 200 or more associations for registered breeds of swine, which maintain records of sires and dams going back for generations.

The associations want people to recognize their animals in the market place, and so, while breeding for better meat production they usually keep as sharp an eye on maintaining the inedible frills of color, design, and the shapes of noses and ears. Every possible combination is represented. There even is a Hereford breed conforming in color and "white-face" pattern to Hereford cattle.

A hundred years ago the goal in domestic breeds (in the United States particularly) was the rounded "lard type" of hog. He was low and short, and they called him "chuffy," but this all ended with the crash of the lard market. Breeders rushed to produce "big-type" hogs that were leggy and long. One breeder during the height of the craze for tall hogs claimed to have a boar "so tall it makes him dizzy to look down."

1). *Tamworth—England. Red. Erect ears, long snout.*

5). *Yorkshire—England. White. Erect ears, short snout.*

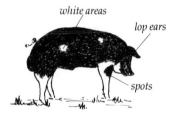

2). *Duroc—U.S. Red, lop ears.*

6). *Chester White—U.S. White. Small, lop ears.*

3). *Poland China—U.S. Mostly black with variable white areas and spots.*

7). *Landrace—Denmark. White. Long bodies, large, flopping ears.*

4). *Berkshire—England.White. Black bodied, with white "points," or extremeties, particularly feet. Lop ears.*

8). *Hampshire—England. Black, with white front legs and shoulder "belt." Erect ears.*

A FIELD GUIDE TO SOME
COMMON PURE BREEDS OF SWINE

This guide is meant to point out distinguishing visible characteristics between eight common registered breeds found in North America. Swine of other breeds occasionally encountered, and hybrids representing combinations of these and other breeds, give us an array of hogs only an expert can hope to tag with breed names.

Summary of Heritability Estimates
for Various Traits in Swine

Traits	Approximate heritability, %	
Productive traits:		
Litter size at birth	5-15	
Litter size at weaning	5-15	Low
Litter weaning weight	10-20	
Feedlot traits:		
Post weaning gain	25-35	Medium
Feed per unit gain	30-40	
Carcass traits:		
Length	50-60	High
Backfat thickness	45-55	
Loin eye area	45-55	
Yield of lean cuts	35-45	

Source: Agriculture Canada.

In turn that craze ended when it was found more money could be made with an animal somewhere between the chuffy toads and the leggy range-runners. This intermediate came to be known as the "meat type" hog, and it is what we have today, balanced in some parts of the United States and particularly in Canada, Great Britain and the Scandinavian countries with the "bacon type"—a longer pig, able to produce larger and leaner bacons than would normally be expected from the shorter "meat types."

Other characteristics—behavioral ones, for instance—find their way into different breeds. They may be good or bad depending on

Hog types.

Fat "chuffy" type   Leggy type   Meat type

management or goals. For instance, mothering ability or an animal's ability to forage on pasture are heritable traits that would be of great or little importance depending on management schemes.

Attitudes Toward Pigs

As pigs have changed or been changed with time and by the influences of different cultures or economic demands, so can you find different attitudes toward pigs and those who keep them. They are probably best known in our world for the bad things about them. They are pushy animals. They seem driven to more than satisfaction of their appetites. Push, push, push and "the Devil take the hindmost," an old saying that must have come from a person familiar with the way pigs will trample anything in their way to try and be first at the trough.

THE DEVIL-PIG

In times past it has been thought the oinker with the curly tail might be an incarnation of the Devil, or at least Devil-possessed. Look between the pig's front toes. There are holes here where the Devil himself must have entered! Actually they're scent glands, but this superstition went on for many years.

Perhaps it is a Devil-link that lies behind a lingering superstition on the coast of Nova Scotia that the mere mention of a pig can bring bad luck at sea. They say there are some older men still fishing today who would turn a boat around and head for the wharf should anyone aboard be so rash as to say the word pig.

One overworn joke that must crop up every year somewhere along the shore—and did the first year we slaughtered here—is the spiking of ears, rooter and tail from a killed pig to bow and stern of a fellow's boat. The more superstitious the victim the louder the hoots of laughter in anticipation of how he'll react at finding his way to a livelihood so corrupted.

I don't know the origin of this superstition, though it may have come here from Scotland where, according to the authors of *The*

Symbolic Pig, "the pig has always been unpopular." They go on to say that, "indeed, abhorrence may not be too strong a word. In Fife and in the Northeast it was believed that the very mention of the animal by name was enough to bring disaster.... It was particularly

The Prodigal Son was a swineherd. From reproduction of engraving by Dürer published in The Symbolic Pig.

unlucky to refer to swine by name while at sea, more especially when baiting lines."

Efforts to avoid calling a pig by name in Scotland and other parts of the world have led to their being referred to as "the short-legged ones," or "the grunting animals," the "grunters," or even this strange one—"the long-nosed general"—which *The Symbolic Pig* says is used in parts of China.

THE ESTEEMED PIG

There have been times and places throughout history in which the pig has been held in high regard. In medieval Europe pigs frequently were the choice for church carvings. They appear playing harps and fiddles, tooting on pipes, being ridden to battle or involved in about any human pastime you can imagine. In those times, too, the Wild Boar ranked with lions and stags as the choice of animal on heraldic family crests. In some cultures the pig, or its images, have been regarded as fertility and good luck symbols. I recently was shown a candy from Germany in the shape of a pig, standing at the base of a horseshoe. The candy wrapper bore a shamrock.

Zeuner says that the people of ancient Crete would not eat pork out of respect for the pig whose squeals drowned the cries of the baby Zeus and so saved him from his father, Cronus, who had vowed to eat the child.

Swineherds, who were absolutely the lowest of the low in ancient Egypt, were given a patron in the Christian church, a saint by the name of Anthony who, according to one tale, put off going to the aid of a prince until he had seen to the welfare of an ailing piglet.

Who knows what political legends lie behind the tales of "The Three Little Pigs," or the rhyme that goes with kids' bare toes? I only know that a lot of us grew up on the side of pigs.

CHAPTER 4

Behavior and Form

Understanding the behavior of an animal species is so basic to the art of management that I wonder why it is often overlooked or only given passing attention. It could be simply more evidence that most animal management books are written with the idea that we all grew up on farms, and so of course we all know... and so on and so forth.

Normal Behavior

Little research has been done to establish the normal patterns of behavior in the pig. We know lots more about cats, rats, and chimps, because these animals have been studied in attempts to find out more about ourselves.

BREED DIFFERENCES

Behavior often is related to form or build. A pig does what it can with the body it has been given. However, among the world's pigs there can be differences of behavior between individuals within a breed as well as differences that seem to hold from one breed to the next. For instance, taking breed differences alone, some pigs are more active than others, the Tamworth being considered "vitalistic," as compared to the Durocs, which were described as "dull and plethoric" in a 1907 study. In 1961 a study found Hampshire boars more virile than Yorkshire or Minnesota No. 1 boars.

Behavioral characteristics may be considered good or bad depending on what a farmer hopes to do with pigs. Tamworths are supposed to be able to manage quite well on their own outdoors. A Landrace behaves well under conditions of confinement and intensive management, but might not have the nesting or foraging instincts sufficiently sharpened for successful rearing outdoors with a minimum of care.

I have been told that the British Saddlebacks were selected, in part, for their ability to do well on pasture. The Hampshire, too, does well outdoors, but sows may lack that degree of docility found, say, in Landrace or Yorkshires, that helps these breeds thrive and reproduce successfully in confinement.

As intensive management and close confinement become more prevalent in commercial swine production, so grows the emphasis on breeding and selecting hogs whose behavior suits these conditions. Anyone thinking of raising hogs in some older-fashioned way should keep this fact in mind and perhaps look for suitable older-fashioned breeds or types.

INTELLIGENCE

I am skeptical of claims that the pig is one of the smartest animals. There are stories about their learning to open gates (as recounted in

A young boar representing a modern breed. Sturdy and trim.

the last chapter) and for a time during the nineteenth century there was a fad among circus performers of teaching pigs to pick up lettered cards to spell words to give the impression that here were "learned" beasts.

Darwin said pigs are highly intelligent, but we don't usually let them live long enough to show it. If he was right, then the pig is unique in my experience as the animal with the greatest lag between physical and psychological development. They're sexually mature at six or seven months, yet still unable to figure out that if they would stand back and let you fill the feed trough there would be enough food to go around, and less on the floor.

Pigs are the fastest-growing farm mammals, and so their demands for food are understandable. But their absolute jam-banging rush to get there first, without the least caution, would seem a hindrance to survival in the wild. No, pigs don't think. They are four-legged embodiments of the "gut reaction."

SKELETON

Pigs have rugged skeletons. Ensminger says they support greater weights, "in proportion to size," than do the skeletons of any other farm animal. The coarseness of bone can vary quite a bit, and breeders are constantly trying to maintain a balance between fineness of bone—which will provide higher percentages of meat on a carcass—and strength of frame. They also watch the overall shape of

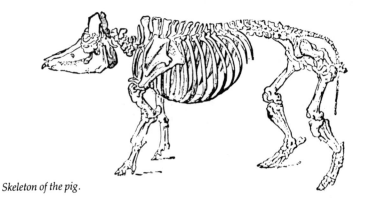

Skeleton of the pig.

the pig. Among changes that can occur here is the length of the body, which often is related to the number of ribs the pig is born with. Longer, "bacon-type" animals usually have pairs of ribs in the upper end of the possible range from 13 to 17. Sometimes a pig will have one more rib on one side than the other, but typically they come in even pairs, with 14 to 15 pairs being the most common.

ROOTING AND SMELLING

From day one, pigs are rooters first. They lead with their noses and are more prone to going under fences rather than through or over. That nose is not at all like any other. It is especially not at all like the human nose, which is so sensitive to the slightest bip. The pig's nose, short or long, ends in a floating disc of cartilage tied to muscles that leave its owner free to move it about like a shovel.

Once a 100-pound pig got loose here and was rooting up the lawn, troweling along, digging a trench about four inches deep at a slow walk. We watched in wonder at the speed and ease of the digging, and laughed to think what was going to happen when the pig reached the summer-baked and packed gravel driveway. Our

From day one, pigs are rooters.

laughing turned to awe to see gravel give way just as the sod had before it—no hesitation, not a sign that the nose had encountered something close to concrete.

Despite their excellent reputations for rooting and smelling—note how they are used to find deep underground truffles, and how the Toomer brothers trained a sow to point game by smell—pigs do seem needful of tasting all manner of things. It is as though they didn't trust their noses to tell them whether or not something might in fact be good to eat.

BITING

A pamphlet I read warns against letting small children get around pigs because they might get eaten, but I do believe that's a bit alarmist. Although pigs *can* bite, they seldom do. I'd imagine children are in greater danger around strange dogs. Some sows are mean and like to nip a nearby leg. But they're in a minority. Most often the orneriness in a sow comes when they are about to farrow or think they have to protect a new litter. Certainly I would watch small children, though, and warn them to stay out of pig pens and not to stick their hands in or hold out fingers that the pigs might wish to taste.

The pig normally bites with a sideways swipe of its head rather than coming at you snout-on as will a dog. The boar also goes at you with a sideways swipe, but there's much worse than a bite in store for the person caught with that pass, especially if it's a boar two years or older whose tusks have never been clipped. The boar potentially is a VERY DANGEROUS ANIMAL. Many experienced handlers will not go into a boar's pen without a hand-held portable hurdle to serve as a shield, and even then only when the boar is occupied with eating or mating. It doesn't have to be an angry boar that rips open a leg or thigh. He could do it in play. Nor should the boar be treated as though he is a wild, vicious beast, prodded with sticks, left in isolation and never handled. "A boar that is frequently handled, petted, given the company of a bred female and led to the altar (regularly) is rarely a problem," writes Gloria Corbett. "Just remember to be cautious and to discourage playfulness."

EYES AND VISION

The pigs' eyesight is not thought to be especially good, but it hasn't been very much studied, any more than has the pigs' ability to see colors. Generally they do have poorly developed muscles of the type needed for sharp focusing. Pigs have those inner eye structures—cones—that in humans are responsible for color perception, but just how the pig perceives light of different wavelengths is not known. Although their ancestors were nocturnal, domestic pigs are the only four-legged farm animals that don't have that inner-eye layer of light-reflecting tissue (*tapetum*) that gives off the beautiful eye shine that deer jackers look for and that scientists think aids night vision.

The position of pigs' eyes on the sides of their heads gives them good lateral vision (when the ears don't interfere). Whereas a cow or horse will stare at you head-on, a pig, like-as-not, will nail you with one squinting eye.

The pig, like as not, will nail you with one squinting eye.

Chewed ears—scars borne by low pig on the totem pole after a squabble with pen-mates.

EARS AND HEARING

Pigs' ears can be small and erect, as in the Yorkshires, or huge and low-flapping as they are in the Landrace, hanging down over the eyes like sun visors. There is every modification between these extremes. Yorkshire ears are perhaps the most noted for the way they stand and quiver when the gilts or sows are in heat and are being mounted.

I believe pigs hear very well. It is almost impossible to sneak into their barn. They are always on the alert for sound from a new quarter, and when they hear it they freeze like rabbits, holding their breaths, waiting for a better reading of what's up. They are reported to like the sound of music. A large piggery in old Mexico many years ago hired boys whose sole job was to sing to the animals.

VOICE

Pigs have many sounds of their own, including a so-called song-of-love (*chant de coeur*) of the boar that recognizes a sow in heat.

Their other voices can be broken down into a vocabulary: There is the warning sniff or noseblow, often followed by a challenging harsh, bark of a snort, very gruff and abrupt: "Harff!"

There are the anguished moans and roars of hungry pigs, and the satisfied grunts—into a trough at last filled with feed.

The quietest and gentlest pig sound comes from a contented sow settled down to nurse her brood, a sound pig farmers anxiously wait to hear from a nervous mother just farrowed and feared liable to savage (kill or maim) her own piglets if she doesn't settle down.

At the other end of the sound spectrum comes the shrieking squeal of a pig being forced to do what it doesn't care to do. It doesn't matter whether you are merely pushing the pig aside or really doing him harm. It's all the same to him. Once that threshhold is crossed he lets you know.

Yes, pigs make many different noises. These sounds probably mean more to other pigs than they do to us. Comstock says constant grunting is a sound that keeps a pig herd together. She also tells of

> ...a certain lady, a lover of animals, who once undertook to talk pig language as best she could imitate it, to two of her sows when they were engaged in eating.
>
> They stopped eating, looked at each other a moment and forthwith began fighting, each evidently attributing the lady's remark to the other, and obviously it was of an uncomplimentary character.

Last, but marvelous, Gloria Corbett, who with her husband John runs a small commerical hog farm in Nova Scotia, reports that pigs can whistle, "sounding deceptively like a human, but with a rather spare, melodic line. I thought we had ghosts or pranksters in the barn until we tracked the whistling down."

FEET AND MOBILITY

Like cows, pigs have *dewclaws*—those small, back toes up high behind cloven hoofs or trotters. But in pigs the dewclaws are relatively larger and longer, and are of great aid to the animal when it finds itself on soft or uneven ground.

Pigs can swim, and by some accounts have been known to enjoy the sport. Ordinarily, though, a deep river, pond or ocean shore will hold the animals back.

Despite their rotund shape and short legs most pigs, especially when they are young, are fast runners and able dodgers. It is no wonder that a greased pig chase, cruel as it has to be for the piglet, has provided such a delight over the years. Pursued, the piglet lowers its head and charges for freedom. If there is a body in the way, tough. Over you go and with little chance of grabbing one of those piston-pumping trotters as he passes. Back in the old days (and maybe today in the southern mountains of the United States where pigs often run free) a way to judge if a pig was properly fat for slaughter was if it could be run to ground and caught by the kids.

Pigs can learn to jump over barriers, so it is important to build the first pen or fence high enough that this skill is never learned in the first place. I failed to do this one year and that pig eventually could scramble over a four-foot board fence. When the pig first showed she could jump I should have added two feet to the walls, maybe more—enough to stop that action right away. Adding a board at a time just continued the problem. This same idea should be kept in mind with fencing any animal: build it right first, so that the fences don't get tested.

SOCIAL BEHAVIOR

Pigs are herd animals, as are their wild cousins, but we seldom notice it because we keep the animals confined. As with other herding or flocking animals, pigs follow that system of bosses and underlings known as "pecking order." A senior boar will lead the herd, but when there is no boar loose with the sows an older and experienced sow rules the sty. Size does not always determine which older pig will be boss. For instance, in my own sty right now the smaller but more aggressive gilt rules over the barrow who, though a littermate, must outweigh his sister by at least 30 pounds.

Pigs can be as absolutely brutal to one of their number low on the pecking order, as can a bunch of chickens bent on pecking a sickly pullet to death. They aren't—as appears to be true with chickens— attracted by the sight of blood. But they will climb into an argument between two pigs, and butt, push and pummel the underling to death. In Chapter 7 (Housing and Handling) there will be several suggestions on how to avoid pig fights.

WALLOWING

Pigs are famous for wallowing. They wallow by choice when the weather is hot, both to cool themselves, and to help thwart external parasites, (flies, mosquitoes, lice or mites). Tanned pigskin is tough, yet the skin while on the pig is hardly more resistant to scratches or bites than our own. Slap a white pig and your hand leaves a red welt. Fly and mosquito bites leave the red spots and blotches of a bad case of the measles. By wallowing, pigs may coat their hides with mud that protects them from bites, and sunburn (which is especially a problem with white pigs) and that may help smother mites.

Pigs are not well equipped to cope with either high or low temperatures. In fact piglets are totally incapable of regulating or maintaining their body temperatures for the first two or three days after

Given the opportunity, pigs display artistry building nests and will literally bury themselves beneath straw to keep warm.

"Dammit, who's hogging the blanket?"

birth. This is why they are given heat lamps or other aids to keep warm. Adult pigs have few sweat glands and most of those are on their snouts. When the weather is hot they must seek a cool spot in which to lie. They seek shade and/or a wallow. If they haven't a wallow they will lie apart from their friends, and with their noses to the wind.

If there is no water or mud for a wallow they sometimes will dig up the ground to get their bellies close to the cooler subsoil. They may urinate to create a mud wallow for heat-, sun- and fly-protection—thus fueling the widely held belief that pigs are the foulest animals going.

Of course, you cannot have this attitude about pigs if you know what they are doing and why. In fact, it is well known among those who have worked with pigs provided with cool, clean surroundings, that they are the cleanest of farm animals. Pigs do not groom themselves. In fact they would be hard-pressed to lick their bodies clean, considering their compactness of form that doesn't allow for much bending or turning of neck or back. This may be why pigs are particular about where they urinate or defecate. Under normal moderately warm and clean conditions, with no leaking water-bowls, no overcrowding, or other causes for confusion, pigs in a pen will choose one spot for a toilet and stick to using it. "Dirty" pigs crop up on occasion, and frequently they are orphans. This lends support to a theory that manuring in one spot is learned behavior.

I have also seen gilts or sows that would stand in the "bathroom" while directing their urine elsewhere—maybe outside the pen or even into the feeding area. This can be annoying, and may be hard to correct. The female pig does urinate more or less forcefully and directedly behind her. The barrows and boars are no problem in this regard as they stand quite still, perhaps treading a bit with their hind legs, while urinating in pulsing dribbles.

When temperatures drop below 50°F. (10°C.) pigs, because of their short and sparse hair covering, begin sleeping close together in whatever sheltered spots they can find. They will gather materials for nesting, as mentioned at the beginning of Chapter 2.

WEATHER FORECASTING

I am reminded of having read in at least two places about pigs as weather forecasters. Youatt says it is an old Wiltshire saying that "Pigs see the wind," meaning they know when a storm is coming. I've never seen it myself nor spoken to anyone who has, but Darwin said, "it is a sure sign of a cold wind when pigs collect straw in their mouths and run about crying loudly. They would carry it to their beds for warmth, and by their calls invite their companions to do the same, and to add to the warmth by numerous bedfellows."

SCRATCHING

Older books on pig management may stress a need to provide scratching posts, the idea being that pigs, incapable as they are of scratching themselves with trotters or snout, need to be able to rub their bodies against a rough log planted in the sty. This need to scratch is aggravated by lice and mange, and is greater when pigs are outside, wallowing in mud and in other ways having to deal with insects and the elements.

The Wild Boars on the Nova Scotia reserve were constantly

scratching themselves. I particularly remember watching one piglet as he sidled up to a rock to scratch one side, then backed around to rub the other, being quite artful at applying rock to itch. But I have not seen this incessant sort of scratching in my own or others' domestic pigs, and if I did, more likely I would first look for bugs and only later for a scratching post. Pen walls and fences should take care of the odd itch.

Form

COAT AND COLOR

The pig's coat is usually straight and smooth, and a curly-haired appearance often is a sign of poor health or a cold environment. In

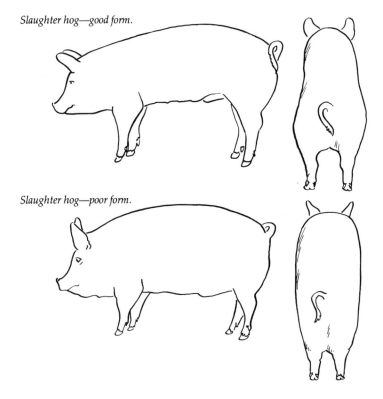

Slaughter hog—good form.

Slaughter hog—poor form.

some few breeds, such as the Mangalita and the Lincoln Curly Coat, the bristles are normally curled. However, I don't know of these breeds being represented in North America.

Pigs come in three basic colors: black, red and white. This is plenty, when you add the variety in which these colors can be mixed to give spotted pigs, saddled pigs, and pigs with horizontal stripes. They may even be tiger striped, though this term is not too precise, since it has been used to describe both tri-colored spotted pigs and bi-colored ones with vertical striping.

Pigs may be called "blue" when they have white bristles on a black skin. They may be called "blue roans" when they have white and black hairs mixed together, or "red roans" if the mixed hairs are red and white. The red color of pigs is said to vary the most in shade—all the way from a light orange or "apricot" to a deep rust.

FREAKS

There are a number of things that can go wrong in the development of a pig. Some are more critical than others, in that they could lead to poor growth or illness, while others might only be considered unsightly. All of the following "freaks," critical or not, should be avoided in breeding stock, since they are heritable traits. I'll put the critical freaks in italic type to indicate those queer piglets that you

Parts of the hog.

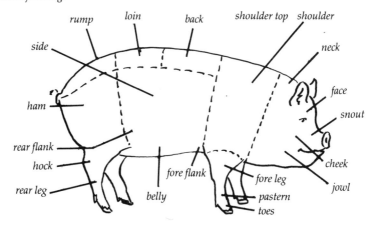

ought to avoid when buying a feeder piglet, since the abnormality
could lead to difficulties or even to loss:

Chin tassels (may be as
    long as three inches)

extra front toes

fused solid hoof (some-
    times called "mule
    footed")

thick front legs

atresia ani (anus sealed
    over at birth, sometimes
    correctable with surgery)

inguinal hernia (called a
    scrotal hernia when it
    appears in the male)

umbilical (navel) hernia (see
    note on hernias, Chapter
    15, Health)

congenital blindness

small or absent external
    ears

harelip

cleft palate

kinky tail

paralyzed hind legs

absence of teats (fewer than
    twelve or fourteen)

blind teats (teats that are
    small or have inverted
    nipples)

mon or *cryptorchidism* (one
    or both testicles unde-
    scended from the boar
    pig's abdomen)

CHAPTER 5

Buying a Piglet

Buying for Pork

Raising a piglet for pork traditionally is a springtime thing—May and June. That way you raise the animal through the summer when there's a minimum need for shelter and when a maximum of food value goes to meat rather than shivers. Come cold weather and an end to flies and your pig is ready for butchering.

However, fall and winter should not be ruled out. Much depends on housing and on what you're feeding. If it is going to be all purchased feeds, garbage or by-products from a constant, year-round source, then season becomes less important. Pigs given a good tight "bedroom" with a door to keep out the cold can do very well in the dead of a northern winter. Likely they would do better than a pig raised through a hot summer climate without provisions for keeping cool.

Even when it may take more food and time to finish a pig in winter there is always the chance that the local market for pork will be better in February than it was in October when everybody else raising a pig has an extra quarter to sell.

Some people raise a summer pig and a winter pig, because, although one pig would supply all the pork they would eat, it is said that pork's freezer life is only five to eight months (see Chapter 17). Our own experience and that of others has been that frozen pork will keep satisfactorily for a year.

45

HOW MANY TO BUY

I've almost always bought two or three piglets at a time because we had the room, and because it's well known that two do better than one as they compete at the feed trough. Time and again you'll see if one gets up to eat the other(s) will be up in a moment too, as if wildly afraid of missing out on a magnificent morsel. The more they eat, the faster they grow.

Then, too, if there's only one pig it's so much easier to get attached to it and to feel more and more unhappy as butchering day draws near.

It takes no more work to raise two or three pigs than one. Too, it used to be figured, even with purchased feeds, that you could raise two pigs, sell one, and get your own "free." But the margin of profit with purchased hog feeds is way down. It would easily take the sale of two or three pigs to cover the cost of the one you keep.

There are strong arguments also for raising only one piglet at a time. The main argument is available food. What if you are an average family discarding a North American average of about two pounds of edible garbage a day? You have a garden from which are available another number of pounds daily of cabbage leaves, bad tomatoes, etc., and maybe you have a cow or a couple of goats, and there's whey from cheesemaking and buttermilk from the churn. Well, take these leftovers, add a bit of pasture, and you've very nearly got one pig's ration. No sense raising two.

WHERE TO BUY

The best place to buy a piglet is from a good neighbor—assuming you have one who raises pigs and sells single piglets or small lots of them. The next best is to follow up an ad in your local paper. There almost always are "piglets for sale" in spring at least. Ask around. See who has a reputation for sound, healthy piglets.

Sometimes it pays to go some distance from home where piglets are selling for less. We have done that. In fact, we once had the good fortune to buy five piglets, three of which we sold on the way home at enough of a profit to more than pay for the trip.

The chanciest place to buy a piglet is a farm auction. There seldom are many if any health requirements for animals put on the block, and you probably won't know where they came from. We have bought at auction and come away satisfied. It was just luck, though. You can't really know just what you are bringing home—what parasites, what diseases.

WHAT TO LOOK FOR

The object of your search—if the goal is pork—is a 30- to 40-pound piglet that has been weaned and on its own at least a couple of weeks. It will be (or should be) six to eight weeks old. A 30-pound pig ten weeks old is a runt, and while it might do an acceptable job of growing from there on, it might also be a slow-poke and food-waster to the last.

Breeds should not matter, they are all so close in performance. Besides, if you are inexperienced, success will hinge mainly on management and feeding. Just buy the best you can from what is available, doing your utmost to see that the choice is a healthy, robust animal.

See that the animal you buy is active and that it has a smooth coat—unless it happens to be one of the rare curly coated breeds. See that the pig is not coughing or sneezing repeatedly and that its eyes are not runny.

Little pigs shouldn't shake or limp when they walk. They shouldn't be skinny, giving them that "razor-back" appearance. Nor should they be flat-sided, giving them the end-on appearance of a loaf of bread.

Watch out for swollen joints or for other swellings on the limbs, particularly those that indicate the presence of abcesses. Also see that they don't have bulges beneath their bellies or in the groin, indicating hernias. Most commonly hernias appear at the belly button of male or female piglets or in the scrotum of the males. They do not necessarily mean trouble if you are only setting out to raise a relatively light slaughter pig. But there is always the chance that, as the pig grows, the hernia will get worse. A loop of intestines could get twisted and blocked within the bulge, causing death.

See that the pig is not anemic. This is easier to detect with white

A mild scrotal hernia on a piglet can grow dangerously and lead to a crippling deformity or death as the animal matures. This unfortunate boar also shows characteristic symptoms of mange, including thickened skin and hairless patches caused by incessant scratching.

pigs. A seriously anemic white one will appear bleached out rather than a healthy pink. To be sure, pinch an ear gently and see if there is a marked delay before the color returns to the white pinch mark. In a colored pig, it will take more experience to recognize the proper degree of pink flush of the eyelids and gums. The gums, too, can be pressed to see how the color returns.

If possible, avoid a pig that has diarrhea or that is constipated. I say "if possible" because you might not be able to tell in a pen of pigs which one has done what. A pig with serious diarrhea (scours) often will show signs of dehydration, such as dull, sunken eyes.

Avoid a pig with drooping head and tail. A pig *does* have a "tell tail," that is up and curled when he's feeling good, and often is drooped when he's not. A drooping tail in itself is not always a sign of poor health—it is not as reliable a sign of sickness as an up and curled tail is that things are pretty much *ok*.

Unfortunately, the tails are commonly removed on young pigs headed for the meat market because, under the crowded conditions of commercial feeding pens, tail-biting may become a problem. It's an unfortunate thing, not only because it's hard to "read" a stub, but also because in a sizeable pig the tail may provide the only decent hand-hold on an otherwise awkward package.

Among piglets of the same litter or age, try to buy the biggest. If you like bacon, try to buy the longest as well. This means being able to look over a whole batch of pigs. This is hard or impossible to do at a sale, where they may only let two or three pigs at a time in the ring, and where the smart seller will have matched and grouped the piglets so you have a hard time telling the runts and wastrels from the profit-makers.

BARROWS & BOARS

Some experiments have indicated that barrows (castrated males) grow faster and fatter than gilts. I haven't found a noticeable difference. I've found them about the same, and so I pay no attention to sex, only making sure that a male piglet has indeed been castrated. Castration avoids the possibility of getting meat that is a bit tainted or "gamy" in flavor.

A boar that has not been "cut" or "changed" will have fairly prominent testicles in a tight scrotum up high beneath the tail and anus. It is not a bag, as it is on a bull or a ram, but more of a shallow pouch held close to the boar's body. And yet it should be obvious enough on a piglet that has not been castrated, unless perhaps the animal is a cryptorchid. (See Chapter 4.)

HOW MUCH TO PAY

The best way to judge the worth of a piglet is to compare its asking price to what it would bring on the commercial market. Commercial growers are continually in the market for healthy pigs, and they have a good idea what they can afford to pay and still realize some sort of profit four months down the line when it comes time to sell to the packer.

There are a number of formulas that commercial growers follow in determining how much a piglet is worth. One calls for paying one half of the wholesale market price for 100 pounds of pork, less $1, for a 30-pound piglet. For piglets over or under 30 pounds, multiply the difference by two-thirds of the market price, and either add or subtract this figure from the base price.

Another formula calls for payment to the weaner-producer of half of the wholesale market price of 100 pounds of pork for the 30-pounder, plus a bonus of a dollar or more if the piglets from that producer have proven out well in the past. Then, as with the previous formula, there will be adjustments in the price for piglets over or under 30 pounds.

Generally, then, you should expect to pay roughly half the market price for 100 pounds of pork for a 30-pound piglet, give or take three or four dollars.

Going back to my earlier statement that the worth of a piglet can be judged by the asking price, you can see that the warning flag should be up if a good-sized piglet is selling for five or ten dollars less than it would on the commercial market.

This is not to say that occasional bargains aren't going to be found in a barn or auction. It could be that cheap pigs are being raised on a small scale by a person who figures a couple of dollars return above the cost of feed is all that is needed to keep going. Maybe this person doesn't want to bother finding a commercial market for the piglets. Maybe commercial buyers don't want to bother with this piddling small producer. You're in luck.

Buying Pigs for Breeding

A thorough discussion of breeding is beyond the aims of this book. However, here are some of the most important ideas to keep in mind.

The selection of breeding stock is a continual process. It should begin with a look at a pig's parents. It does not end until the sow or boar is culled from the herd.

PARENTS

Initial selection should be from among piglets from better-than-average parents. They should be large, robust and healthy. It is essential that no evidence of hernias is seen in either the gilt or the boar piglet, and that they *both* have at least six pairs of evenly spaced teats, the spacing being important to allow room for the piglets to nurse. Some breeders are now selecting for at least eight pairs of teats, but this only makes sense with longer pigs, since a short-bodied sow can only accommodate so many piglets.

The number of teats varies greatly, and it is a highly heritable characteristic carried by both sexes. It is also essential that the young boar have two decended testicles, because testicles that do not descend do not produce viable sperm.

FORM

The overall pig destined for the breeding farm should be long-bodied, with little neck, and a smoothly rounded back that curves around to form a deep, full ham. From front or rear the body should be fully rounded rather than pointy, razor-backed, or flat-sided. The underside of the sow should be quite straight, not sagging. The boar's underline should be high and curving—a line that is just about parallel with his back.

Legs must be strong rather than fine. The joints should not be bulging. The pasterns should be strong and upright, but not ramrod straight. Some "spring" is essential if the animals are to be kept on hard floors. The toes on any hoof should be equal in length.

Pigs for breeding should be the best from the best litters. The reason for saying this is because the best of a good family is more likely to pass on those good traits than is an exceptional piglet from an unexceptional litter.

Breeders should come from farms where management is along the lines of what you provide. It may be a mistake to buy fancy-looking animals from a farm where they were pampered, if you are not going to be providing the same sort of home.

Many important characteristics cannot be read in a weaning piglet. A wonderful-looking weaner from excellent parents still could develop into a below-average adult—could, at 200 pounds live weight, carry two inches of backfat and have weak hind legs. For this reason people building up their first breeding herds probably would be better off buying pigs—particularly boars—near breeding age.

THE BOAR

You should expect to pay more for a boar than a sow because of his single influence over the offspring of many sows. It makes sense to have one that is registered purebred and, better still, one that has been officially tested for growth and backfat. This is what titles like ROP (Record Of Performance) mean. Some governments assist in the purchase of these boars as their way of supporting improvement within the industry.

Unfortunately, this still is no guarantee of breeding performance. An exceptionally good looking, purebred and ROP-tested boar will not necessarily pass along what it has to its offspring. Or it may pass it on only when it is bred to certain other animals.

A boar that consistently passes on certain traits to its offspring is said to be "pre-potent" for those characteristics. When a boar and sow do breed well together—throwing exceptional offspring time after time—they are said to "nick" well.

HYBRID PIGS

Usually, crossing two pure breeds of swine produces piglets that grow faster and in other ways out-perform purebred piglets. This response to cross-breeding is called *hybrid vigor*, or *heterosis*. Some farms produce feeder pigs that are results of three-way crosses. First, simple, cross-bred sows are produced from two pure breeds. Then these sows are bred to a third pure breed to produce commercial feeder pigs. Heterosis is said to be at work even more in these three-way-cross piglets.

I haven't scratched the surface of what there is to know about breeding pigs. Anyone seriously interested in the subject should turn to one of the newer books listed in the bibliography and talk to agricultural college and extension service representatives. For up-to-date addresses of U.S. Swine Breed Associations write: USDA-ARS, National Staff Program, Room 302, Building 005, Beltsville Agricultural Research Center—West Beltsville, Maryland 20705. In Canada, write: Canadian Livestock Records, Ottawa, Ontario.

CHAPTER 6

Handling Pigs

CATCHING AND TRANSPORTING PIGLETS

"Don't buy a pig in a poke," the old saying goes, by one account coming from old Greece when under Mohammedan rule, and when piglets were sold in the dark of night to avoid upsetting the bosses. Upset was the buyer who got home and found he had bought a cat.

People say don't even carry a pig in a poke (bag). Why? Is it cruel? Unsafe? I don't think it is either, unless it is a very hot day. In most cases, and for hauls taking no more than an hour or two, a burlap feed bag with the dust and dirt beaten out makes an ideal pig rig.

In the dark of a bag, a piglet soon quiets down. It gets plenty of air through the open weave of the burlap and, best of all, a pig in a burlap poke is secure. A hastily slapped-together box might not hold a pig, and the biggest danger and possible trauma to a young piglet will be if it gets away, jumps into traffic and so on. Each 30-40 pound pig should have its own bag. Tie the throat of the bag tightly with bailing twine or heavy string. If you use plastic feed bags, be sure to cut holes for air.

A bag is not good in hot weather for more than a few minutes. If you must use one longer, sprinkle it with water. The evaporation will be cooling. Heat is hard on a piglet already overheated from the stresses of being chased and caught by its heels. For this reason it is far better on a hot day to use a wooden box or cage that allows plenty of ventilation. Keep the piglet in shade. Sprinkle him with water if he's looking flushed and is breathing rapidly.

The best way to catch a piglet free of its box or poke is by a back

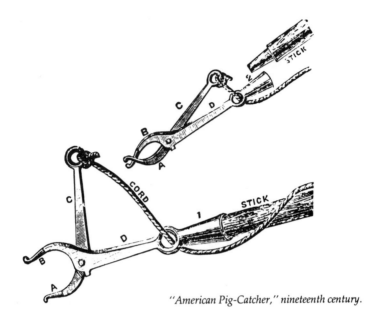

"American Pig-Catcher," nineteenth century.

leg, above the hock. Don't pick one up by its ears or tail, no matter how small. Some people do this, but it is hard on the little guys.

No matter how you pick up a piglet it likely will shriek bloody murder. Held by a hind leg it usually quiets down shortly. If one doesn't and you can't stand the racket, cradle it in your arm and gently but firmly muzzle it with your free hand.

Picking up a small pig to give it a shot in a fanny muscle or for some routine examination is a one-person job if you swing the piglet between your legs. Slight knee pressure behind the piglet's head keeps him in place. Other techniques for handling piglets are explained in Chapter 13.

TRANSPORTING LARGE PIGS

Large pigs are best transported in a crate or the box on a pick-up truck. Crates are better for single animals because they can be built just wide enough to take the pig while not allowing it to turn around or to fall down should the truck turn too fast. Ensminger suggests a layer of wet sand—an inch or so—be laid down under a large pig being shipped on a hot day.

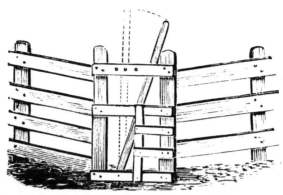

A pig-catching gate can be used in a straight fence or in a cul-de-sac. A pig is held by the movable lever as soon as it sticks its head through the opening.

A pig will more readily enter a passageway than a box, and so the best crates give this illusion either by having their far ends merely barred with mesh wire or by having doors that can be dropped or swung into place after the animal has entered. Hungry and docile pigs often can be coaxed into the darkest crate with a bit of food.

A crate with handles can be carried to the pen for loading, thereby avoiding any need to drive the pig over open ground. If the loaded crate is too heavy to lift, push it on log rollers, using a couple of planks for ramps wherever necessary.

Last summer we experimented with a floorless crate that we dropped over a sow in order to move her across the yard. That way we didn't have to lift her, the crate being used simply as a guide. It worked reasonably well, but it would have been better built with smooth sides—to keep the pig from lifting the crate with her snout —and with a secure top to keep her from attempting to jump out.

Moving Pigs on the Farm

CALLING

With patience, pigs may be taught to come when they are called. They learn to respond to a blast on the swineherd's horn, and special bagpipes were once used by swineherds in southern Italy. In keeping with the old one-liner,"Call me anything but late to supper," pigs have been called many things over the years. "Zuck... zuck," in *Lorna Doone;* "Soo-o-oo-boy," in another book; and "Soo-ey .... Soo-ey," from my own childhood recollections. I wonder if this last didn't come from the pigs' scientific family name of *Suidae.*

HERDING

Swineherds called their pigs and drove them, too, with dogs and whips. Youatt found this description of a German swineherd's whip: It was

> ...one of those terrific whips which must be seen to be imagined. At the end of a short handle, turning on a swivel, there was a lash about nine feet long, formed like the vertebrae of a snake, each joint being an iron ring, which, decreasing in size, was closely connected with its neighbor by a band of hard greasy leather. The pliability, the weight and the force of this iron whip rendered it an argument which the obstinacy even of the pig was unable to resist.

DRIVING

In *The Family Cow* I talked a bit about "flight distance," which is the space an animal likes to keep between itself and some threat. Ignoring flight distance—call it crowding—makes herding an awful job. Crowded animals either balk or bolt when what you want is slow, even progress.

Flight distance varies with the species and with the nature of the threat. Between a fairly docile pig and a human stranger standing quietly, the distance is about five feet. If the stranger is moving toward the pig, the distance easily doubles.

When herding a pig, move toward the animal slowly, noting the point at which it begins to retreat. Nothing will be gained by trying to get any closer, because you have already crossed that flight distance line that spells security.

If you are positioned to head a pig off at an alley or pen gate, stand well back so that the animal does not balk and run. Don't crowd the pig until it turns to investigate in the direction you wish it to go.

A hand-held hurdle that presents the appearance of a solid barrier is a great aid in herding pigs. It may be of plywood or any other light material, about 3 feet square. Watch swine judging at a fair or exhibition to see how expert handlers keep hogs under control in wide open spaces with nothing more than hurdles. On the farm, some prefer a more sophisticated hurdle—double in width and hinged in the middle.

Sometimes a rope around a pig's hind leg will be enough restraint and provide enough steering wheel to direct a pig on a walk. Or you may find walking beside the pig with one hand on its tail and the other on an ear will work.

Braced knees pushing from behind are a tremendous help when it comes to persuading a balky pig to move down a narrow alley or into a pen or crate. Those who say the way to move a pig is to simply cram a bucket over its head are being misleading. Sure, a pig backs up if you put its head in bucket. But that's only a part of the problem. You've got to direct the backing or you're nowhere. So if you try the bucket trick, have someone else guide the backing by hauling on the pig's tail. It sounds easier than it may be. Sometimes you have to literally pick up that pig's hind end in order to shift him in the right direction. And you have to move fast to keep the bucket crammed over the animal's head.

Tails are so helpful. It is a shame when they get removed. The Corbetts routinely dock tails on pigs going to hog feeding operations but leave them long on any they plan to keep for breeding. Moving day for sows on their farm provides some wonderful sights. A sow is released from her pen or stall and is directed—piloted—completely from the rear by means of her tail. "Come on, let's go, girl," says Gloria, giving the tail a hike, and off they go down alleyways and

Two hands for beginners? Gloria Corbett is no beginner, but "steering" a large sow by the tail can take both hands on a sharp turn.

around corners with the driver calmly coming behind, lifting and shifting the sow's rear end whenever necessary, always lifting opposite to the direction the sow is to go.

If you have to drive a pig onto a truck, try to get the back of the truck as nearly level with the pen or barn floor as possible. It's good to think of this when you're building a barn or loading area for pigs—and other animals,too. The next best arrangement is a long and only gradually sloping ramp or stairway, with high boards on either side and built only wide enough for a pig to pass and not to turn around.

Portable loading ramp.

ESCAPED PIGS

At some time or other on most farms pigs escape. I'm getting better and better at retrieving them. It comes from practice, of which there is lots here, since I'm prone to leaving gates unhooked.

We used to get quite upset when we discovered a pig loose, feeling sure, I guess, that the animal would soon be off running free in the woods never to return. We would scramble after those pigs, and they didn't need any grease to make them impossible to stop.

Now we have come to realize that if your fortune is ours and you're well away from a busy highway and anxious neighbors—and if all gardens are well fenced—there is little to be concerned about with a pig on the run. Especially this is so when the pig has spent a month or more on your own place. That barn or pen is home, and it is unlikely he will go far, unless in fact you *do* go chasing and yelling about. More than likely a little taste of feed will bring the pig into the pen. If it does not come right away, wait a while. Hunger tames a pig. Show it a scoop of food, and once it has shown an interest, slowly walk backward into the pasture or pen.

RESTRAINING PIGS

Big, tame pigs are suckers for a belly rub and often can be encouraged to lie down for a scratch, at which point wounds or other minor problems may be tended. At other times there may be a large pig that so loves to eat that, come meal time, he will chomp away without objecting to examination or routine treatment.

   A noose. One way to restrain the largest hog is with a noose or "hog-holder" over its snout. A piece of small diameter (¼-inch or less) rope or cable, or a purchased or homemade snare may be used. The idea is to slip the noose far back in the pig's mouth, behind the tusks, and over the top of its snout, and pull tight. The noose tightens as the pig draws back. It must be very uncomfortable, because most pigs will not draw back hard. They just stand and shriek—fortunately don't think to charge their tormentor.

Pig snares.

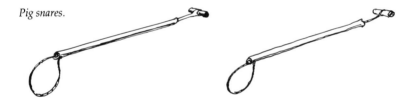

   It takes little strength to hold a feeder pig this way. A safer bet with a large hog is to tie the free end of the noose to a solid post. If there is difficulty getting the pig to take the snare, smear the rope or wire with a bit of molasses or the like. Greed overcomes discretion.

   Hog-holder. A hog-holder works on the same principle as the noose. But in this case the tool is a two-foot long bar of iron with holes at either end. One hole is larger than the other, one being for pigs, the other for hogs. A holder with flat, hexagonal holes is recommended in *Restraint of Animals* over ones that are angled and that have holes that are round or triangular. The authors say these last "may take the skin completely off the pig's nose."

Snares or narrow-gauge rope looped around the top of the snout behind the tusks immobilize a pig.

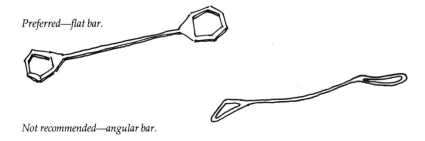

Preferred—flat bar.

Not recommended—angular bar.

Although a noose of small-diameter rope can be used in place of a snare or holder to snub a pig, it is not as good because it takes time to loosen. The snare or holder lets go as soon as you relax pressure.

Snubbing should not be overdone. It does hurt the pig, and too, once a pig has been snubbed it may be difficult to catch it that way again.

A large pig that will not lie down for a belly rub or allow itself to be snubbed may have to be cornered and upended for treatments or examinations. Try this method, again from *Restraint of Animals:*

Make a large, slip-knotted loop and flip it over the pig's head. As the pig moves to get away, swing a wide open jump-rope loop of your rope's free end in its path. As soon as the pig's front feet have stepped over the jump-rope throw a half-hitch in the free end and pull tight. The pig is in harness. Force the pig against a wall and have someone hold the rope. To drop the pig, reach under its belly, grab both far legs and pull toward you. *Jerk* is probably a better

Make sure that the loop is well back in the pig's mouth.

description. You want to catch the pig by surprise and have it down before it knows what is up. A pig can be kept down by straddling its body and holding its front feet in the air.

Very large hogs, of 300 pounds or more, could be injured if they were dropped to the ground with the suddenness that goes with jerking the off legs from under them. You might try snubbing the snout with a rope and then running the free end of the rope back, inside one back leg and around, forming a loop around the hock. Pulling the rope pulls the back leg toward the hog's head and he falls—on a good heap of bedding if you have properly set the stage.

CHAPTER 7

Penning a Pig

Indoors or Out?

The dozens of ways to pen a pig illustrate the pig's great adaptability to conditions imposed by people acting through the pressures of tradition and economics.

We know that pigs need protection from extremes of weather. They can't stand to be too cold or too hot. Periodically they can stand to get wet, but only when it's hot. Otherwise evaporation, enhanced by wind or drafts, will cool their bodies off too much.

Whether we provide that necessary protection by giving the pig a little house in a large pasture, a little house with a pen attached, (or tie the pig to the little house with a leash) or some fraction of space in a large barn is a question of economics and resources. If you have no buildings but lots of pasture, consider the first route. If you have no land but a big barn that could easily be remodeled, consider the last.

In North America, the trend in commercial farming has been away from raising pigs on pasture or range toward *confinement* rearing in environmentally controlled sheds. These are workable systems wherever labor and land costs are high and energy costs are relatively low, because they allow one or two people to raise many hundreds of pigs each year.

BUILD IT RIGHT—NOT TOO SMALL

Three mistakes often made in building homes for feeder piglets stem from misjudging their incredible rates of growth. First, houses and

A typical and economical set-up for keeping a couple of feeder pigs.

pens are built too small. Second, not high enough. Third, houses
and pens are built before buying, with visions of big pigs running
through our heads. The results are enclosures with such wide gaps
they don't hold the 30-pound weaner.

Building too small is the most cruel mistake. Give the little guys
plenty of room and they'll grow into it very soon. Figure at least a
five by five foot floor area for sleeping for a piglet to be taken to
market weight, with another 10 square feet of sleeping area for each
additional pig. Generally you should allow twice these amounts of
space for living—for feeding, manuring and exercising—in confinement.

If piglets come home before their proper house is built, a good
temporary shelter for one or two in warm weather is a 50-gallon
drum with the top cut out by acetylene torch or a cold chisel. Lay the
drum on its side, brace it from rolling with stones or logs, and throw
in some hay for bedding.

Another idea for daytime care of small piglets is a homemade,
low-slung garden cart sort of affair—a four-by-four or so box with-

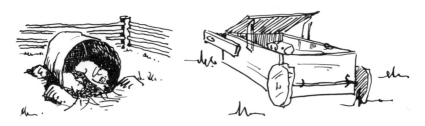

Requirements for the Accommocation of Swine

Item	Sows	Pigs- under 50 lbs.	Pigs- 50 to 200 lbs.
Feed Lot- hard surfaced	25 sq. ft. per sow	8 sq. ft. per pig	20 sq. ft. per pig
pasture	1 acre per 2 sows with litters	1 acre per 25 pigs	1 acre per 10 pigs
Building pen area	25 sq. ft. per sow under 400 lb. 35 sq. ft. per sow over 400 lb. 64 sq. ft. per sow under 400 lb. with litter 80 sq. ft. per sow over 400 lb. with litter	3 sq. ft. per pig	8 sq. ft. per pig
pen partition height	3 ft.	2 ft. 8 in.	2 ft. 8 in.
Slotted Floors- pen area per pig			7 sq. ft.
-slotted floor area			25-100 percent
-slot width			5/8 in.-1 in.
-slat width			1½ in.-5 in.
Self Feeder Length	6 in. per sow	2 in. per pig	3 in. per pig
Feed Trough Length	1 ft. 6 in. per sow	10 in. per pig	1 ft. 1 in. per pig
Individual Feeding Stall	1 ft. 6 in. ×6 ft. 6 in. ×3 ft. 6 in. high		1 ft. 1 in. ×5 ft. ×2 ft. 6 in. high
Farrowing Stalls- 5 ft. ×7 ft. (including 2 creeps) clearance under creep partition	9 in.		
Water	1 watering cup per 15 sows	1 watering cup per 25 pigs	1 watering cup per 20 pigs
Feed	1 ton per year	1000 lb. feed from birth to market	
Bedding	½ ton of straw or equivalent		

Source: Agriculture Canada.

out a bottom, covered at one end, and equipped with wheels and handles for moving about the yard.

A HOUSE AND PEN

The most common arrangement for keeping pigs on a small scale is a sty divided into a small house and an exercise area. Even if the pigs are being kept in a large barn, an individually roofed and enclosed "bedroom" may be helpful in cold weather.

It takes less material to enclose a square area, but though it is more costly, a long and narrow rectangular pen is better with pigs because they will keep it cleaner. Maybe it helps to keep them oriented, having feeding and manuring areas more or less taking up far and opposite ends of a pen. But there is such a thing as *too long* in a pen. In fact, anything over twelve feet may be wasted space, since this is the average distance, Ensminger says, that a pig will travel from the feeding area before defecating.

Fences, walls or hurdles enclosing small areas must be at least four feet high, because some pigs (especially when they are young) become good jumpers. Many publications suggest three feet is good enough, but I have had 100-pound pigs that could scramble over that low a wall with no trouble at all. For larger, outdoor enclosures, fences three feet high are usually adequate, because the animals are not as anxious to get out.

Inside

FLOORS: WOOD OR CONCRETE?

The floor under an indoor pig area may be concrete or wood—solid or slatted to let the manure go through. Combinations are possible, for instance, with solid flooring under most of a pen and a slatted floor in a lower, what might be called "manuring," area.

Concrete is preferred by people who want a floor that will last almost indefinitely and that is easiest to clean. Wood is preferred for its warmth and by those who feel that it is easier on the legs and feet of heavier hogs.

The best indoor pens I have seen were in Missouri on a farm

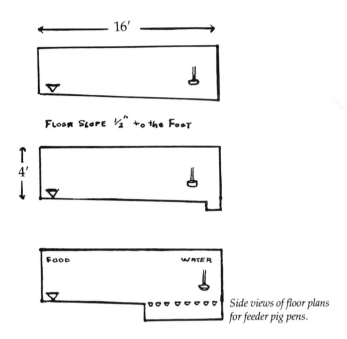

Side views of floor plans
for feeder pig pens.

producing three-way-cross feeder pigs. The floors were concrete, about two-thirds solid, and sloping to the manuring area. The remaining third of the total was slatted and had a manure cistern beneath. The next best had floors that were entirely solid concrete sloping gently to the bottom edge of the pens where, in turn, a manure drop sloped away to a covered lagoon. In both cases there was a minimum of labor needed to keep the pen clean. The floors were sloped the recommended half inch to the foot.

My own pens have to be shoveled out. They are quite crude, have wooden floors and wooden feeding troughs. But they always have a low spot or section away from the feed trough that automatically becomes the manuring area.

Cleaning up after two or three pigs is no chore though, taking hardly five minutes a day. Therefore, I cannot see much advantage in having slatted floors or self-emptying manure drops for the tiny raise-a-bacon kind of operation.

A solid wood floor should be as smooth as possible for ease of cleaning, because it is less likely to get worked on by pigs following their rooting instincts, and because it will be easiest on the pigs' feet.

The cold of concrete floors beneath bedding areas may be alleviated by incorporating a layer of insulating material beneath the top inch or so of cement. Angus Rouse, swine technician with the N.S. Department of Agriculture, reports success laying the surface concrete over a layer of fiber egg flats.

Concrete floors can easily be *too smooth* for safe walking. To avoid this, when putting down a concrete floor, either make shallow grooves an eighth- to a quarter-inch deep in the surface just before the concrete sets, or trowel the surface off with a piece of rough, unfinished lumber.

WALLS AND PARTITIONS

Indoor pen walls or partitions between pens may be of pipe, heavy planks, poles, poured concrete or concrete blocks. Wood, though it is not suggested for permanent housing because it gets chewed, is what I use because it is what I have. Maybe some day we will go to pipe or reinforced concrete. We certainly would do so if we were raising pigs on a commercial scale, because wood rots, gets ingrained with dirt and eventually gets chewed away wherever an animal decides to "crib." When building pens or fences with old painted lumber, turn the painted sides out, so they can't get chewed. Many old paints were lead-based and deadly poisonous.

Poured and reinforced concrete walls three to eight inches thick may be filled with rocks or—for their insulating quality while saving concrete—with empty, inverted quart juice cans. This idea, from the Corbett farm, takes a bit of dexterity and patience. You have to pour the concrete slowly, with an extra hand to make sure the cans don't float out of position.

DOORS

A good deal of thought should go into the hanging of a door on a pigpen, because at least once (when the inhabitants go to market or slaughter) and perhaps many times, a door that swings out, in, left or right may be a great help in directing traffic. Swinging doors are great, and the most versatile is a swinging door hinged both sides with removable hinge pins so that it may be swung from either side.

LIGHTING

There are people who claim pigs fatten faster in darkness, and that when it's dark there are fewer fights among crowded animals. This goes back years to Youatt, who said that "sleeping rooms" should be dark, since animals fatten more rapidly when they lie down and sleep between meals than when they wander about. So you will see hog barns without a single window. Inside is a completely controlled environment, probably totally dependent on electricity to keep air moving, and to give light when it is needed at chore time.

By another argument the best barn is one the operator and employees are happy to be working in, and from that standpoint I'll have all the windows winter heating costs will allow.

One caution is to watch for rickets in pigs being kept in dark or glassed-in places. It is likely they will need a dietary source of vitamin D, the "sunshine" vitamin.

TOTAL CONFINEMENT OF SOWS

It is becoming more and more common on large commercial farms to keep sows totally confined for several months of each year through-

Sows are confined to stalls in this barn, held by neck yokes (left) and belly straps. Belly straps allow pigs more freedom, obviously.

out their lives. They are kept in crates or in narrow stalls in which they may be held either by neck collars or belly straps. Neck clamps have to be bought especially for the job. Belly straps may be special hog belts or they may be seat belts out of General Motors junkers or those of any other manufacture that release by pressing a recessed button that a neighboring pig's snout can't operate.

Neck clamps are quite snug on a pig. Belly straps are allowed to run a bit slack and do seem to give the pig more freedom. They are commonly attached by means of short lengths of chain to eye bolts recessed in the floor. Neck clamps may be fastened by chains running to both sides at the front of the stall.

Because pigs cannot see behind themselves, they are reluctant to back up. Most will not step back and over a pipe or other barrier a couple of inches below tail height, so only that is needed to keep them in stalls or crates that are blocked in front with feeders and so forth.

The floor under any stall or crate that totally confines a pig should

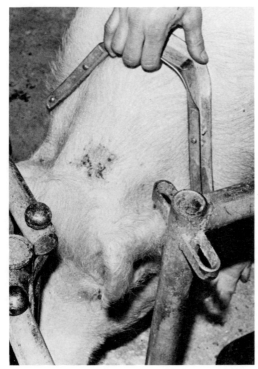

Neck yokes can be too tight.
Watch out for open sores.

slope both to the head and tail ends from a point beneath the shoulders. Otherwise water or feed slops—or worst of all, water from a jammed fountain or automatic bowl—will soak the entire stall.

VENTILATION

You run into ventilation problems when you fill a barn with pigs. A sow gives off about a gallon of moisture a day through breathing. Add evaporation from urine and spilled water to this and you can see why it's important to get engineering help in designing buildings for concentrations of pigs. You need to create an environment of constantly renewed air, yet without drafts. Books and other publications dealing fully with these problems are listed in the Appendix.

Speaking of fresh air, one old and musty system for raising pigs that I've come across numerous times is to keep them in the manure pit below or beside the cow barn. It's a winter system, really, related to the western practice of letting feeder pigs "follow" beef steers. The pigs root through the manure, gleaning grains and other nutrients, including important B vitamins.

Outside

A PIG HUTCH

An outdoor hutch for feeder pigs can be a simple, "A" frame affair. A floor is necessary except in the driest climates. Otherwise the pigs root out a nest beneath their shelter and this becomes a wallow when it rains.

More substantial hutches are illustrated. Some can be used the year around in coldest weather, for feeders, for sows dry or farrow-

Outside hutch banked for winter.

ing. In wet, cold weather several hutches can be drawn together and banked high with hay or straw. They can be placed on a concrete platform for the wet, cold season and supplied with lights for doing the chores on winter mornings, and with heat lamps for new litters.

As an added note regarding raising pigs in winter, here is a quotation from *Behavior of Domestic Animals* that favors the practice: "Pigs reared outdoors in huts or with outdoor runs grow faster, are healthier and eat more than winter pigs reared indoors." How much of the added food goes to keeping warm is not revealed.

Well-started weanling pigs and older animals do not need supplemental heat in winter if they are provided with small houses (or bedrooms if their pens are in large, drafty barns) with doors, into which they can retreat between meals.

A doorway on a pig house or bedroom should be three feet high and two feet wide—wider for pregnant sows group-housed, to prevent injuries from crowding. Height is critical, because a pig can't scrunch down and wriggle through a low opening the way a dog or cat can, and a low lintel will likely scratch and bruise their backs. Sills should be no more than four inches high.

If doors are installed on a house or bedroom they should be hinged at the top so that they automatically swing closed. Sometimes where groups of pigs are being housed there will be "in" and "out" doors.

HOT WEATHER

All pigs, but especially white ones because they are more susceptible to sun burn and heat stroke, must be given shade if they are

A simple 'A' frame pig house. Floors are a must in all but the driest climates.

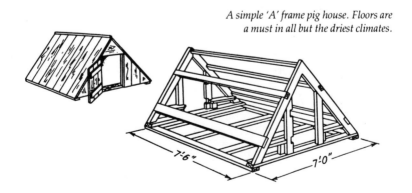

7'-6"

7'-0"

Effect of Air Temperature
on Rate of Gain with Swine

Mean live weight (lb)	AVERAGE DAILY GAIN (LB/PIG) AT AIR TEMPERATURES (F) OF						
	40°	50°	60°	70°	80°	90°	100°
100		1.37	1.58	2.00	1.97	1.40	.39
150	1.27	1.47	1.75	2.16	1.82	1.14	— .19
200	1.19	1.57	1.91	2.22	1.67	.88	− .77
250	1.10	1.67	2.08	2.14	1.51	62	−1.36
300	1.02	1.77	2.24	2.06	1.36	.36	−1.95
350	.94	1.87	2.41	1.98	1.21	.10	−2.53

Source: Agriculture Canada

outdoors in clear, hot weather. If pigs aren't using their hutch in hot weather, maybe the door should be removed and a window cut in the back to provide ventilation.

White or aluminum paint or whitewash (though it's less permanent), will lower the temperature of a hog house, according to one source "by as much as 15 degrees Fahrenheit." This could be important in purely economic terms. Ensminger says growth gains are small when the temperature goes to 90°F. and above—that at 60°F. you can get 100 pounds of gain from 400 pounds of feed but that at 85°F. it will take 1200 pounds for the 100 pounds of gain.

A final note for people using old barns or sties for their pigs: Before putting the pigs in, go around with a hammer and saw and

Effect of Air Temperature
on Feed Conversion with Swine

Live weight (lb)	FEED CONVERSION (LB FEED/LB GAIN) AT AIR TEMPERATURES OF						
	40°	50°	60°	70°	80°	90°	100°
70 to 144	4.9	4.4	3.6	2.5	12.7	7.3	8.3
166 to 260	10.5	5.1	3.6	4.1	4.2	12.0	

Source: Agriculture Canada

get rid of any old nails sticking out of the floors or walls, and of any jagged pieces of wood that might injure your animals. Puncture wounds can lead to abcesses or even tetanus.

YARDS AND PASTURES

It is hard to say how big a sod- or dirt-floored exercise area for feeder pigs should be. It depends on the number of pigs and on whether or not, and how often, they may be moved to new ground. The area should not become a mire of mud and feces, and really the only way to avoid at least some part of the yard becoming just that is to move the pigs with regularity. If the pigs must be kept in one area they

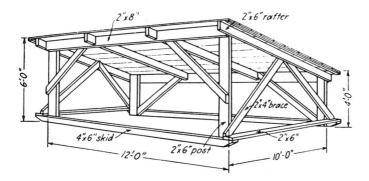

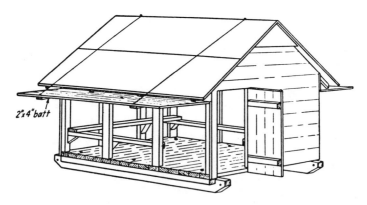

Pig hutches should be moveable. A simple frame with roof (above) may suffice for warm weather, while a four-sided hutch with hinged sides, floor and door (below) is good for winter. Both are built on skids.

should have at least thirty square feet of yard per animal. Also, they would be better off if they were given a platform on which to stand while feeding. Friends have built a successful hutch with this in mind, providing a balcony along the front of the hutch on which the feed and watering troughs are placed.

To avoid a buildup of parasites, pigs should not be run on the same ground year after year. For this reason there is nothing to be gained from building a permanent pig house that's joined to a muddy little lot of land. Pig hutches should be built to be moved, and that means not building too big. If it is too big to be picked up it must be built on skids so that it can be moved with a tractor, horse or some other vehicle.

Portable hurdles of wood or wood and wire may be used for outdoor pens. They may be tied together or joined by reinforcing rods run down through eye bolts at their ends. I have heard of using freight pallets for hurdles, which sounds excellent if the price is right.

FENCING

When fencing pigs, either in a barn or outdoor paddock or pasture, the thing to remember is that they are rooting animals and more likely to go under than over or through a barricade. The bottom has to be best. It has to be low, at ground level or even below.

Electric fences on stakes 10 to 12 feet apart are excellent for pasturing feeder pigs or sows, and ought to be feasible wherever there aren't laws against them. Two strands of barbed wire usually are recommended, though "common" wire will do. One strand may be sufficient on level ground and where the vegetation is clipped back, so that it is sure not to touch and ground-out the system. Jerry Belanger recommends two strands and that each be hooked up independently to batteries (or whatever power sources are used) so that there is less chance of having the entire system grounded or otherwise put on the fritz.

When two strands are used, one should be about six inches off the ground and the other a foot to 18 inches above the first. Steven Thomas suggests using lengths of reinforcing rod for posts on an electric fence, with "J" bolts to hold the insulators to which the wires are attached. If you're looking for economy, wooden stakes will do, and instead of buying porcelain or plastic insulators, try folds of heavy plastic or slices of old tire casing nailed like belt loops to the posts.

Any fence that is not electrified must extend at least to ground level. Barbed wire (12 gauge) should be spaced at three-inch intervals. Heavy (11 gauge or less) page wire fences should be joined by at least one strand of barbed wire along the bottom, unless the pigs have been ringed to prevent their digging. Special hog-type page wire, two feet high, makes an excellent fence when stretched tightly and topped with two strands of barbed wire six and twelve inches above the top of the paging. Posts should be six to eight feet apart.

Board or pole fences are good, but unless a person has a supply of either, the cost of materials may be as high as with page wire and ten times the work. Also they make the least portable kinds of fences, which is something to consider when rotation of pasture is so important with hogs.

MIXING ANIMALS

Pigs should not be pastured with milk cows, says Spencer. "Very occasionally a sow or even young pigs will acquire the very bad habit of sucking the cows." Usually, he says, they go to cows that are lying down though one instance was reported where "the cow stood to nurse a pig." Teats or quarters of the cows' udders may be injured, leading to mastitis.

There also is potential danger for pigs around a cow with calf. Last year a cow, calf and heifers ganged up on a sow that got loose here. They bowled her over at one point before she could be rescued. I have not heard or read anything against letting pigs run with sheep. However, I have heard more than once that pigs may kill and eat chickens. On one farm where chickens were being allowed to roam free they got in the habit of reaching through the board fence of the pigpen to snatch food from the trough—a habit with fatal consequences for slower birds. Another danger in allowing hens to run with hogs lies in the possible spread of diseases such as TB.

SPRINKLERS & WALLOWS

Sometimes in extremely hot weather it will be good to give the pigs a water sprinkler. While it would be a friendly gesture to let pigs go into a pond, stream, or swamp, there likely would be problems in terms of erosion of banks, and parasites such as stomach worms would be in heaven, being able to deposit their eggs in warm oozy mud. In the past some farms have provided concrete "wallows."

Hot weather comfort—concrete wallow and shade.

These wouldn't give pigs their protective layer of mud but they would be cooling and, in the long run, probably cheaper to operate than sprinklers.

TETHERING

Sows or a small number of feeder pigs may also be kept outdoors on tethers, as are used in England and elsewhere. Sometimes a sow is tethered to a shelter and her brood is allowed to run free. Apparently they do not run far. Some pigs take to tethering more readily than others. From my own experiments with feeder pigs I think it is best to have the tether attached to the underside of the belly strap on a harness. The tether on a large pig has to be made of light chain, for

Some pigs take to tethering.

rope will get chewed. At least one swivel is needed to keep a chain or
rope from twisting into a ball.

PENS FOR MATURE HOGS

Relatively light fences will hold feeder pigs or sows in a pasture,
compared to what is needed to keep sows or boars in a small area.
Boars especially require heavy fencing because they grow so large
and potentially dangerous. In many instances the only thing that
keeps them penned is ignorance. Keep them stupid.

Crates are too confining for boars. Build their fences—and those
for the mature sows being confined—four to five feet high. They
may be of concrete, heavy planks, poles, pipes or barbed wire (again
three-inch spacing) or of page wire bottomed with a strand of
barbed. Fence posts holding wire should be no more than six feet
apart. Poles or planks should be spaced close enough together to
prevent the hogs from getting their noses through to chew upon
them.

As to space for a sow or boar, one recommendation is for a sty that
is a minimum of six by eight feet, connected to a court that is "as
wide and twice as long."

CHAPTER 8

Furniture and Utensils

Troughs

The standard idea is that a pig eats out of a trough, but this doesn't have to be so. Pigs with well-established manuring areas and being hand-fed dry feeds don't mind using a relatively clean, dry piece of floor for a platter.

When troughs are used for wet or dry feeding, allow ten to fifteen inches of feeder space per pig if they are being *hand-fed*—meaning their food is dished out at certain times of the day rather than having it available at all times, as is the case with self-feeders. One feeding station on a self-feeder hopper will serve up to five pigs. Hinged drop doors over self-feeder stations keep weather and rodents out. Piglets three to five weeks old will be able to operate these covers.

An open trough for hand feeding should be divided by cross-bars, both to keep pigs from sleeping inside and to make it harder for more agressive pigs to bully others away from their food. Although it takes up a lot of room in a pen, some are equipped with individual feeding stalls to keep the bullying down. Going a step further, the stalls may be equipped with doors or a pipe that can be dropped behind the animals at feeding times to keep some pigs from gobbling their own food and then chasing neighbors out of their own sections of trough. This last system is better in theory. Too often in practice one sow holds back from entering a stall, or two persist in fighting their way into one until it is too late to drop the gate; another sow has finished her meal, has backed out and is intent on running yet another sister from the table.

79

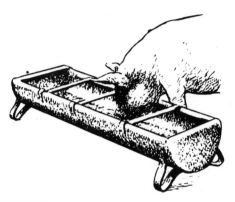

TROUGH MATERIALS

Troughs may be made of wood—simple "V" troughs being common and good—or you can use an old water heater tank cut in half lengthwise, adding legs that can be bolted down. Lengths of pipe can be welded in to serve as dividers. Wooden troughs are impossible to clean thoroughly, though this may not be a problem unless you run into diseases that force complete sterilization of pens and utensils. Also, a wooden trough is usually so light it has to be bolted or spiked down—the same with most metal troughs. Concrete troughs are both easy to clean and heavy. Even if they aren't built in as part of a pen floor, they can be built heavy enough to resist removal and destruction by bored pigs.

A trough mounted with one end protruding from the pen allows for easier watering or slop feeding. Or it may be built beneath a chute that introduces the feed from outside of the pen.

Pigs or sows outdoors may be fed from troughs set on concrete or plank platforms that prevent the feeding area from becoming a bog.

Exercise used to be considered important, especially for the well-being of pregnant sows. I agree, and repeat an old suggestion that, if

Wooden trough filled using a chute.

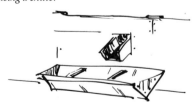

possible, the sows being kept outside should have their feeding areas or troughs across the yard from their sleeping quarters, so that they are forced to walk at least a bit every day.

WATERING SYSTEMS

Pigs may be fed water in their food troughs or they may be given special drinking troughs. They may be watered from buckets held for them twice a day, or automatic fountains or bowls may be installed. Pigs on self-feeders should have water available at all times. One automatic bowl or fountain will be enough for ten to a dozen feeder pigs. Bowls or fountains should be installed at the lower, manuring areas of pens. That way any leaks will not drip

Concrete trough or watering bowl for single hog.

Bucket weighted with sand
3" piece of pipe or hose to
provide a drain.

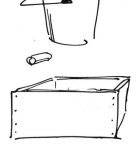

Wooden box form

After pouring 2" concrete floor in box, insert bucket
and pipe and fill box with concrete.

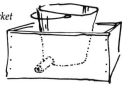

Allow to set, then remove bucket.
Trough will slip out of greased form.
Put wooden plug in
drain hole.

A wet food carrier.

through the lengths of pens. Also the inevitable moist area that goes with an automatic waterer will tend to reinforce the notion that "here" is the pig's place to relieve itself.

Safety and Comfort

CREEPS

Whenever sows are allowed to run with their piglets, the little ones have to be given areas where they can get feed but Mom can't. An area like that is called a *creep,* and there are any number of ways to build one. There should be enough space within the creep for all of the little ones to scramble about for their supplemental feeds. Some creeps are built to accommodate piglets from more than one litter. Entryways eight to nine inches high will keep big pigs out. Sows

Providing corner creep area in farrowing pen.

may get trapped and injured trying to force their way through higher openings.

BEDDING

Pigs aren't always provided with bedding, but when they are, they seem to enjoy it. There are two reasons for using it: one is to provide a soft and warmer nesting area and the other is to sop up urine that otherwise might be wasted.

We use old hay or hay rejected by the cows to bed pigs' sleeping areas, and shavings or coarse sawdust for the manuring areas. Moldy, dusty or wet hay should be avoided in all areas. Some people say you shouldn't use hardwood sawdust or shavings for bedding pigs because it will "burn" them, but we haven't noticed any problems or differences between soft and hardwood products. I've also heard that some people don't like sawdust because it gets in the pigs' ears and makes them shake their heads constantly. But again, we've never noticed any such problems, and in fact I'd wonder if constant shaking wasn't due to ear mites or some nervous disorder perhaps brought on by a vitamin or mineral deficiency.

Sawdust is preferable to shavings in creep and farrowing areas for the protection of piglets' knees. It stays put, whereas shavings tend to get kicked and blown aside. Avoid fine, band-saw sawdust, as it may be too dusty for the health of the piglets' lungs. Bedding in creep areas should be changed frequently.

When the weather is cold—unless barns are heated—pigs love to get a pile of hay or straw to chew up and push into a high pile of a bed. Sows, too, will like having something to build a nest with, but here you have to be careful. More is said about nesting materials for expectant mothers in Chapter 13.

One of the advantages in the use of slatted floors is that bedding is not required. And, in fact, bedding can defeat the purpose of a slatted floor if it accumulates and blocks the slots through which the manure should fall.

CHAPTER 9

The Meat Producers: Digestion and Physiology

There are three things that together make the pig the best meat-producing domestic animal. Pigs are fast-growing; they make the greatest gains in live weight for the amounts of food eaten; and they yield the highest percentages of carcass to their live weights.

Weight Gain

The average piglet weighs 3 pounds at birth, doubles that weight in its first week, and goes on to a finish weight of 220 pounds in less than half a year.

Actual growth rates vary with individual pigs. Litter runts may always be slow. There also may be variations in rates between the breeds. However, growth does not always correlate with a conversion of food into valuable muscle. As an example, *Behaviour of Domestic Animals* says that the Duroc grows faster than the Poland China but less of the nutrients go into building muscle.

You could not compare animals of different stages of maturity this way because, as pigs grow older, fewer nutrients go to building muscle (protein) and more go to the laying on of fat. Fat is a richer substance, needing more energy in its manufacture. *The 1939 Year-*

book of Agriculture (USDA) reports studies showing that it took 1,625 calories to add a pound of live weight to a 75-pound pig, and 2,600 calories to add that pound to a 225-pound animal.

FEED CONVERSION

Comparing pigs and steers as feed converters, Fishwick pointed to studies that showed a 200-pound pig eating 6 pounds a day of dry matter having 4½ pounds of *starch equivalent* (a measure of a food's energy value used in England) will put on 1½ pounds of live weight daily, whereas a thousand-pound (comparable weight) steer eating 24 pounds of dry matter having 13 pounds of starch equivalent puts on 2 pounds a day. "A poor result, in comparison to the pig," the author added.

Pigs yield a carcass that is 69 to 75 percent of its live weight—the difference depending mostly on whether the head, kidneys and leaf lard are included in the carcass weight. Compare this to the 60-percent-of-live-weight-carcass yield in the average beef animal, and to the 50- to 55 percent yield in fat lambs.

When comparing pigs to cud-chewing sheep and cattle it is important to remember that the ruminants are able to do a far better job of making meat out of low-value, hard-to-digest grasses and the like, while the fastest-growing porkers are eating pounds of grains that could build human bodies directly.

Still, the pig shows many improvements over the Wild Boar,

Average Percentages of Certain Parts of the Hog Carcass Classified According to Live Weight

Average Wt., (live)	218 lbs.	289 lbs.
Average Wt., (chilled carcass)	177.5 lbs.	238.5 lbs.
Hams %	17.5	17
Loins %	11.6	11
Bacons %	11.4	11.9
Shoulders %	17	16.7
Head %	8.9	8.5
Cutting fat		
(Back fat, leaf fat trimmings) %	19.5	21.5

Source: USDA *Yearbook of Agriculture, 1939.*

having become a marvelous machine trimmed to those essentials needed for eating and growing. He shows what is called "refinement" over the wild boar ancestor, with his tubular body, short legs, almost no neck, barely any tail, no loose folds of skin and little hair. It takes protein to build hair, skin and long legs—protein that in the domestic pig can go to building meat.

LONGER INTESTINES

There is one important place where the domestic hog has more of something than his wild ancestor—something that should be played up and capitalized on by all who work with swine. That more is intestines. Ensminger says the length of the wild boar's intestines is nine times his body length. Intestines in the domestic pig average fourteen times the animal's body length. Hot dog!—because the longer gut improves the pigs' potential for digesting plant material.

"The pig is usually considered to be omniverous," writes Kenneth Hill in *Dukes' Physiology*, "but under domestication it is essentially herbiverous, and there is considerable microbial breakdown of plant material in the large intestines."

The large numbers of microbes and their breakdown of plant matter within the gut is the secret of the success of cows, goats, sheep and horses in living on these otherwise hard-to-digest foods. The host animal gets the benefit of carbohydrates and fat-related compounds released through microbial digestion, and then gets proteins as well from the bodies of yesterday's microbes gone to their rewards.

There is little of this microbial breakdown in the gut of a carnivore. Even if it could take place, there just isn't that much room. The cat's intestines, for comparison, are only four times its body length. But here comes the domestic pig with a much longer gut than its wild ancestor, and not only is there "considerable" breakdown of plant material in the large intestines, but, Dr. Hill writes, "appreciable quantities of volatile fatty acids are produced (through "gut fermentation") in the cecum (appendix) of the pig...," and that, "Lactic acid and volatile fatty acids are produced in the stomach of the horse, pig and rabbit as a result of bacterial fermentation."

Lactic acid and other fatty acids, together with glycerine, make up the required fats in a pig's diet. The advantage to a pig getting them through gut fermentation rather than through fats added to a ration isn't known. But it would seem that the domestic pig may have more ability to use plant foods, including some that are quite high in fiber, than he is ordinarily given credit for. This would be especially true of older pigs, those reaching maturity and over 100 pounds in weight, since their rates of growth are slowing and their demands for proteins and minerals are slackening off.

The wonderful thing is the adaptable nature of the pig, both inside and out. In spite of this longer, plant-digestion gut, he also can grow—through our own type of enzyme digestion—on a diet that is almost exclusively of animal products. It's done in Newfoundland, where pigs from fifty pounds on may be raised on fish and fish parts. The pig's body takes what proteins it needs for its

own protein-based tissues and breaks the rest down to provide energy a grain-fed cousin would get from plant oils and/or carbohydrates. (See Chapter 12.)

DIGESTION

Whether it is plant or animal food being consumed, digestion begins with *ingestion,* or eating. Pigs, cows and horses take in liquids the same way, by sucking them up, except that pigs are noisier.

Cows have long tongues that they use to gather in grass or grain. The pig's tongue is narrow and too short for gathering. Instead it uses its lower lip, pointed and nearly prehensile, to work foods into position for picking up between front teeth. From there, a lot of head tossing may go into throwing morsels of food back to where they can be mashed between the beveled surfaces of sharp molars.

The front teeth or incisors are pushed forward in the pig. Also, many pigs' jaws don't quite mesh, because the lowers are longer than the uppers. For these reasons a pig may be hard pressed to take a bite out of something large and solid. This could be the reason why pigs may balk at eating a whole squash or pumpkin and yet go right at it if it's ground or crushed up.

A piglet is born with eight tiny but sharp "needle" teeth corresponding to baby (or temporary) tusks and corner incisors. There are two on each side of the upper jaw and two on each side below. On most farms the tops of these teeth are clipped off to prevent injuries to the sow's teats. (See Chapter 7.)

Within days more temporary teeth appear, until by the age of six weeks the piglet has pre-molars and a full set of incisors. By eighteen months the hog has a full complement of forty-four permanent teeth, and only the tusks on the boar, uppers and lowers, continue to grow.

At one time it was thought that boars could be aged by the length and curl of their tusks, but there is too much variation between breeds and with condition and management. Too, nobody caring for the health of hogs (or themselves) would leave a boar's tusks unclipped today. Tusks that appear from beneath the curl of a boar's lips don't tell age so much as they tell of a job that needs doing.

As foods are chewed, they are mixed with saliva containing the enzyme ptyalin (the *p* is silent), that begins changing starches into sugars. We have ptyalin in our saliva, too, but no other domestic animal besides the pig has it in quantities important to digestion.

Foods chewed and lubricated are swallowed into the pig's stomach, a simple, one-chambered stomach very much like our own. Here the food is mixed with gastric juices that go to work on the proteins.

After a thorough churning, the mass of partially digested food (chyme) moves on to the small intestines where it is met by doses of enzymes from the pancreas and bile from the gall bladder attached to the liver. Each of these chemicals has a specific part to play,

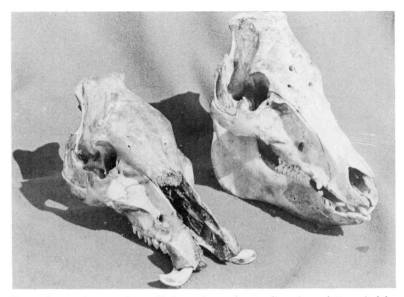

Hog teeth come in many forms. Tusks, molars and protruding nippers better suited for shoveling in the earth than for taking bites out of firm fruits or vegetables. The skull on the left is that of a young Wild Boar.

making possible the breakdown of foods into products small enough for absorption through the walls of the small intestines into the blood and lymph systems.

The large intestines, bigger around but shorter than the small, are left with the job of drawing water from the remains of the chyme. If the walls of the large bowel are inflamed or irritated through poisoning or infection, or have become coated with oil, the pig comes down with scours. Very simply, water is not being absorbed—a serious situation, since any living body needs water for the transport of nutrients.

Carcass Quality

One class of nutrients, the fats, needs more mention here because of direct relationships between the quality and quantity of fats eaten and the quality of the carcass produced.

A slice of bacon shows the way fat is deposited in layers between streaks or bands of muscle. Fatten a pig, and the biggest difference to the eye will be thicker layers of fat and not, as in good beef cattle, an obvious increase in "marbling," or a laying down of fat cells within the muscles themselves. Therefore, an overly fat hog will show wide bands and blankets of fat throughout the carcass, to which many people object.

Whereas proteins and carbohydrates—almost without exception—are chemically taken apart in digestion and later reassembled within the pig's body in new forms, many fats are taken in unchanged. They only need to be broken into tiny droplets of their original forms to be absorbed and deposited intact in the pig's body. In this way fish fat may be incorporated in the pig's body unchanged in form or flavor. The same is true with many other plant and animal fats or oils.

"SOFT" AND "HARD" PORK

The texture of pork, too, can be related to fat in the diet. Lots of soft, unsaturated fats—those that are liquid at room temperature—become soft fat in the hog, making a carcass that is floppy, with sides of bacon that drip fat if they are left in a warm room.

A list of foods blamed for building "soft pork" includes beech nuts, acorns, soybeans (but not soybean oil meal), peanuts and peanut oil meal, corn hominy feed (if more than half of the grain ration), corn germ meal, rice bran (if it makes up more than a third of the total ration), rice polishings, linseed and coconut cake (also called oil meals), fish meal with high oil content, cod liver oil, and in many instances a garbage diet, particularly if the garbage carries a high percentage of unsaturated fats. Rice grain and barley are noted for producing "hard" pork.

Sometimes economics dictates the use of more than a proper amount of soft-pork feed. Apparently this is all right if, toward the last of the feeding period, the pig's diet is changed to one that ordinarily would produce firm fat.

From what people have told me and from accounts in books, it seems that a month to six weeks is all the time needed to change the consistency—or the flavor—of a pig's fat. This is the time Newfoundland fishermen give their fish-fed pigs at the end, switching them to a diet that is mostly barley, to overcome any "fishiness" in the flesh before doing them in. Foxfire tells of bringing pigs off the mountains, where they have been gorging on acorns, to change their fat with corn.

Ensminger says the change of fat consistency that can be accomplished

> takes place more rapidly if the animals are first fasted for a period before the change in ration is made. This practice is called "hardening off." Thus, many hogs that for practical reasons are finished primarily on such feeds as soybeans, peanuts or garbage, are hardened off with a ration of corn or some other suitable grain.

THE IDEAL CARCASS

Management and heredity come into play as a farmer works to produce pork carcasses that are uniform in their weights and amounts of muscle and backfat. Today, the desired carcass—the one that packers want and also the one that probably represents the best return to the commercial farmer (because high-priced feeds haven't been wasted producing excess fat)—weighs 170 to 190 pounds (from a live weight of 215 to 250). It carries approximately an inch of backfat, and has a loin "eye" area about five inches square. (The backfat measurement represents an average of two or more readings along the animal's back. See Chapter 7.)

To produce this pig in 150 days or less, the grower relies heavily on the right genes, since carcass quality is about 60 percent heritable and 40 percent due to feeding and care. Getting those genes takes a selection of dams and sires that are better than average—that have, for example, considerably less than an inch of backfat at five to six months of age.

CHAPTER 10

Pig Nutrition

Many, if not most, stories about the strange eating habits of pigs, can be explained by nutritional needs. Last year I heard about a man who bought a nearly full-grown pig cheap because the animal was failing. It could hardly walk, yet the day after arriving at its new home it managed to escape and make its way to a pit heaped with wood ashes where a highway crew had been burning brush. "And you know? That pig cleaned it up," the astounded new owner reported. "Must have been a bushel of ashes gone when we came on the pig down there." Within a week the pig was a going concern.

In another instance, a family raising some pigs on a cheap mix of ground grains full of hulls ran out of its regular feed and was forced to use a bag of a complete and balanced hog grower. Suddenly the pigs ate and ate and ate. "You couldn't fill them up with that hog grower," they complained. "It's no damned good."

I suspect in both cases that these porkers knew better than their owners what *they* needed to keep healthy and growing. In the case of the wood ashes no doubt it was minerals the sick pig was after, and perhaps all that it needed to get well. In the second case, the pigs may well have been starved for proteins, for minerals, for vitamins or for all three, which they were not getting in adequate quantities from the cheap mixture. What is more, once offered a proper ration, they could eat their true nutritional fill without their systems becoming stuffed with useless, indigestible fibers that made up too much of the cheap feed.

What I am saying is that pigs aren't as blindly piggy as they may at first seem. They do a fair job of balancing their own diets if they are given a chance. It's done this way sometimes when people have lots of corn to feed. They'll give the pigs their free choice of corn in one bin, a high-protein supplement in another, and perhaps a third and separate bin for vitamins and minerals—sort of a cafeteria system.

Such a system does not always work out as well from an economic standpoint, especially if you put an exceptionally palatable protein supplement next to a not-so-tasty grain. Nonetheless, the pigs will stay healthy.

Nutritional Needs

The nutritional needs of pigs can be divided into six categories, or classes. These are *water, carbohydrates* (starches, sugars and digestible fibers), *fats, amino acids* (for building proteins), *vitamins* and *minerals*.

Some of these pigs need daily, some almost daily, and some they can store away so that they may be able to go weeks or even months between inclusions of this or that nutrient in the diet. Comparing body building with house building has always helped me to understand the basic uses and interrelationships of the various nutrients.

WATER

Water can be compared to those roller conveyers at a building site that carry building materials in and the scraps and leftovers away. Water is the most important need because it is the solvent or carrier for all the rest, as well as the medium for heat exchange.

Without oxygen or water there cannot be life as we know it. It's so obvious. And yet a lack of water is probably responsible for millions of dollars lost each year in terms of lowered health and lowered growth rates in animals.

There is nothing difficult about filling a pig's water requirements. If you give it all the water it wants twice a day it will do fine. If it is on a self-feeder it will prefer to have water on hand all the time. Automatic feeders and waterers aren't new to pig farming. Youatt found reference to a hog farmer who had set up hoppers for feeding beans ad-lib "and he having drawn a rivulet of water through each sty, the daily trouble of waiting on the hogs is saved." The date was 1677.

Behavior of Domestic Animals says the average water-to-dry-feed ratio required by all swine is 3:1 by weight. This ratio will vary a lot with the temperature. I have found that it does not take a hot day to double a pig's thirst. Requirements are highest for lactating sows— as high as ten gallons a day, according to Fishwick.

We have found you can leave a bucket with a pig while it is actively drinking or alternating eating and drinking—as it likes to do when it is being hand-fed a dry meal—but that buckets left any longer get demolished. Cheap plastic buckets from restaurants or meat markets may last longer than expensive metal buckets.

Approximate Daily Feed and Water Consumption

Pig weight pounds	Feed consumed, pounds	Water Consumed	
		Pounds	Gallons
50	2.7	5.1	0.61
100	4.7	8.9	1.07
150	6.3	12.0	1.45
200	7.6	14.4	1.73
250	8.2	15.6	1.88

Source: Agriculture Canada

Pigs being fed moistened feeds, fresh greens or wet (even soupy) garbage will require less water. In fact hogs being fed slops may be getting all of the liquid they need—the same with pigs getting lots of milk or related dairy fluids. However, even these pigs should be offered fresh water from time to time, especially in hot weather.

FATS AND CARBOHYDRATES

Back at the building site again, there are the carbohydrates and the fats that can be looked on as the fuels (energy) needed by the tools and workers erecting the pig.

Fats (called oils if they are liquid at room temperature) are 2.25 times higher in energy value than the carbohydrates. As a fuel source they could be natural gas as compared to wet wood.

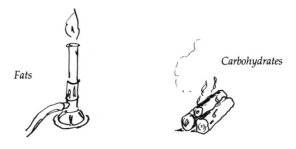

Fats

Carbohydrates

Fats and carbohydrates are not completely interchangeable as sources of energy in a pig's diet. One reason is that fats may be more difficult to digest. Yet a diet that has no fat at all is not good for a pig, in part because many vitamins and other factors needed for the proper working of the whole pig are either carried by fats or are incorporated in fats or fat-like compounds. A pig denied fats or oils may develop dry skin and a rough coat, which may be only the visible signs of worse disorders inside. About 10 percent of fat is a good level in the diet of a growing pig. Any more may result in a fatty carcass, especially if it is fed toward the end of the period of fast growth.

As was mentioned earlier, some fats may be deposited in a pig's body without being completely broken down in digestion. This is why fish oils may flavor a pig unless they are taken from the diet six weeks or more before slaughter.

A low-fat diet may produce a low-fat pig with lean cuts (chops, hams, etc.) that most of us would consider dry and less palatable than fattier models. I have heard of this happening with pigs raised on skim milk or buttermilk, where they were fed milk far in excess of their daily requirements in amino acids.

The main fuel in pig-building rations is the carbohydrate starch, which is found primarily in plant seeds and roots. A kernel of corn is nearly 70 percent starch grains packed together beneath a protective fiber skin and surrounding a high-protein and high-fat germ.

Other carbohydrates are the sugars in the saps or roots, stems, leaves and fruits, and the cellulose and related fibers (of varying digestibility) that are found throughout plants.

AMINO ACIDS

Amino acids, minerals and vitamins are the building materials of the animal under construction. Although all three are vital, the amino acids, often referred to as "the building blocks of proteins," are needed in greatest quantity.

Amino acids build proteins and proteins build the cells of plants and animals. In animals even the cell walls are made of protein. In plants the walls are of cellulose, and this is the main reason why a package of animal material is generally much higher in protein than an equal amount of plant material.

Protein is not a precise word. If houses were animals, the walls would be one protein, ceilings another, and so on, with each major portion of the building being made out of a different protein. The materials to make these different portions would be the amino acids—boards, tarpaper, shingles, bricks, tin, nails or whatever.

There are many different amino acids, just as there are many different materials that go into building walls, floors, roofs, foundations and the like. We can get along without some of the amino acids by making do with others. It is the same with building, where we could erect a wall of brick or of board. But there are certain amino acids—about ten of them in pig nutrition—that are indispensable, or "essential," because no others can be used in their places.

For the sake of illustration, you could think of nails as an "essential" ingredient in house building because their place or role can't be filled with boards, brick or paper. Nor, practically speaking, can boards or other wall-building materials be replaced by stacks of nails.

In considering the protein sources a pig will use to build its own muscular frame, it is important to know not only the *quantity* and *digestibility* but also *quality,* in terms of amino acid content—how many and how much? A "complete" protein contains all ten essential amino acids. One that is complete *and balanced* for pigs will provide all ten essential amino acids in the right proportions for rapid body growth and development. No single plant protein passes the test, though soybeans and peanuts come very close. Certainly soybeans are better for people or pigs as sources of protein than are any of the cereal grains, which are seriously deficient in two or more essential amino acids.

Usually a balance of the amino acids is accomplished by feeding growing stock a mixture that contains some plant and some animal protein. Young growing plants, especially those of the legume family (peas, beans, alfalfa, clover) also are excellent sources of proteins, supplying most if not all essential amino acids—though perhaps not in quantities (or concentrations) sufficient to sustain rapid growth in a young pig.

This need to provide not only enough but the right kinds and balance of proteins is the big difference between feeding cattle (or sheep or goats—the ruminants) and feeding hogs or chickens (or humans). The ruminants don't need complete proteins because microorganisms within their digestive systems build all of the amino acids for them out of whatever source of chemically combined nitrogen they can find—nitrogen being the critical ingredient present in and distinguishing amino acids and related compounds from the carbohydrates and the fats. This is why many ruminants beyond infancy can be fed such a chemical form of nitrogen as urea as a "protein source" when the same chemical would poison a pig.

When a pig is fed more protein than it needs the excess is broken down and used for energy. Proteins are about equal in energy value to the less fibrous carbohydrates. Usually they cost more than sugars or starches, though, so that feeding more protein than is needed may not be economical.

VITAMINS AND MINERALS

The importance of vitamins and minerals in animal feeding was realized near the turn of this century following discoveries that animals fed ideal quantities of foods (supplying all the necessary amounts of carbohydrates and amino acids) still might fail. There were unknown "factors" that could be obtained by putting the animals on pasture or by giving them small quantities of this or that supplement.

Growing pigs need minerals.

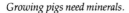

The "factors" turned out to be vitamins, vitamin precursors (chemicals needed by an animal to build a vitamin) and minerals. Most of them have now been identified, and yet you still hear and read about suspected "factors" in some diets that somehow make them superior to others. The search goes on. It is often a slow and painstaking one because the quantities involved are usually minute.

Vitamin or mineral deficiencies show up quickly and dramatically in pigs because they grow so fast. Across North America deficiencies in iron, calcium and phosphorus probably are the most common among the minerals. The vitamins most often lacking are A and D. None of these is costly or hard to supply.

Iron deficiency is a problem with newborn piglets deprived of clean earth in which they might root about and get all of this mineral they need.

Calcium and phosphorus are involved in many chemical reactions and combinations in the body, but most noticeably in the makeup of skeletons. Bones are 90 percent calcium and phosphorus—heavy on the calcium. A growing animal that is building bones, or a sow in late gestation or nursing a litter, will require more of these minerals and a higher proportion of calcium to phosphorus than is generally

Daily Calcium and Phosphorus Requirements of Hogs

	CALCIUM		PHOSPHORUS	
CONDITION AND WEIGHT OF PIG (POUNDS)	Dry ration (Percent)	Amount (Grams)	Dry ration (Percent)	Amount (Grams)
Growing pig:				
50	0.4	2.5	0.3	1.8
150	.3	3.0	.25	2.5
250	.2	3.5	.2	3.1
Pregnant gilt:				
250, early	.25	4.0	.2	3.5
250, late	.4	7.0	.3	5.0
Lactating sow: 400	.45	10.0	.35	8.0

Reprinted from USDA *Yearbook of Agriculture, 1939.*

needed by a dry sow or boar. The recommended ratio of phosphorus to calcium is 1:1−1.5 by weight.

Cereal grains supply phosphorus but are seriously lacking in calcium. Many of the leafy forage crops provide calcium—the legumes and the weed known as *Lambs-quarters* being good plant sources of this mineral. Forage crops may also provide phosphorus if they are grown on soils that are well supplied with this mineral.

A pig that is not fed enough calcium or phosphorus gets rickets because bones don't develop properly. It limps; it gets swollen joints, bent legs, a large head; and finally it becomes paralyzed, stops eating and dies. This happens frequently when attempts are made to raise pigs on all-grain diets, mill "sweepings" or most types of garbage, whey or other rations that aren't supplemented with minerals.

Rickets can also develop in pigs getting plenty of calcium and phosphorus but not enough vitamin D, which is needed to put these two minerals to work. Enough vitamin D can be made right in the skin of a pig exposed to an hour or two daily of direct daylight (glass filters out the active ultraviolet rays). Indoor pigs will need to have vitamin D in their food. Sun-cured hays, especially the leafy legumes, fish oils or irradiated yeast, are usual sources.

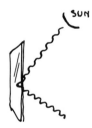

*Glass blocks vitamin
D-producing light waves.*

Vitamin A or its precursor, carotene, comes from yellow foods—yellow corn, carrots, sweet potatoes—or from green forages where the yellow is hidden by chlorophyll. Well-cured and well-stored hay, particularly alfalfa or the other leafy crops, also are good sources of the vitamin.

Pigs whose diets are lacking in vitamin A are subject to colds and eye problems, including night blindness, and to breeding and farrowing difficulties.

Rations lacking vitamin A may be corrected with fresh green herbage, * alfalfa, fish oils (except for feeder pigs on their last weeks to market lest their flesh be tainted) or one of the factory-produced forms of the vitamin.

Further mention of vitamin and mineral requirements may be found in the Appendix. For a more complete handling of their places and importance in swine feeding read Ensminger, Morrison, and materials from the National Research Council (also listed in the Appendix). Home-grown feeds may reflect local trace mineral deficiencies in the soils. Avoid nasty surprises by consulting local authorities about these possibilities.

---

*Give pigs an hour or two daily in fair pasture and they will solve both their vitamin A and D requirements. Other trace elements of their diet—some known and some yet to be discovered—will likely be supplied as well. The pigs will get some exercise for muscle development and will in all ways be healthier for their moments out of close confinement and in the fresh air.

CHAPTER 11

Feeds and Feeding

Using pre-mixed balanced feeds from the store makes a pig the simplest animal to feed. A canary couldn't be easier. Certainly not chickens, cows, goats or horses, because with pigs, feeding can be a matter of filling a self-feeding hopper once or twice a week with the purchased dry rations and seeing that they have water—no seed cakes or cuttlefish bones, salt blocks or hay. It's very neat, and I think there are lots of pigs raised each year that would do better if they got this kind of management and feeding.

Instead, attempts are made to cut corners and feed bills—which, granted, are high with commercial pre-mixed rations—by substituting cheap feeds that are high in fiber and low in nutritional value.

We shouldn't have problems feeding pigs properly, though, because their needs are so like our own. In fact a pig fed out of the kitchen pots and pans of a healthy, well-nourished family would do well. The major differences are a pig's ability to produce its own vitamin C—saving on breakfast juices there—and the pig's enormous appetite, which is only matched by an enormous rate of growth.

Pre-Mixed Rations

Basically there are three different commercial pre-mixed *balanced* rations for feeder pigs that provide all of a pig's daily nutritional needs. These rations differ the most in the amounts of protein they provide. *Pre-starter* (20–22 percent protein) is fed early-weaned piglets (under five weeks) beginning as early as the first week and carried on for three to five, by which time they have been slowly switched over to a *Starter* (having up to 20 percent of protein). Starters may be continued until the piglets have reached 65 to 70 pounds. Then they are switched again, this time to a *Grower* (providing about 16 percent protein). The *Grower* is continued to slaughter.

PRE-STARTERS AND ANTIBIOTICS

Pre-starters are introduced to piglets that are still nursing. An allowance of 16 pounds of Pre-starter per piglet should be ample. Only the smallest amounts are used initially, because the piglets only toy with it those first few days. Although it may not be eaten, the Pre-starter must be changed daily to keep it as fresh and attractive as possible.

Most commercial Pre-starters are medicated. Antibiotics have been added to discourage disease and promote growth. The routine use of these medicated feeds should be discouraged, and especially it needs not be part of a program in a small-scale operation where animals usually are less crowded; where sickness can be discovered and treated as it crops up, and where it matters little if a pig takes a few more days to reach a good size for slaughter.

The "beneficial" growth-stimulating properies of the antibiotics and other chemicals were first discovered in the early 1950s. They were an instant success in the marketplace, and before long every commercial farmer was just about forced into using them to stay competitive. After 20 years, however, some of the dangers began to be suspected. In 1971 an Agriculture Canada bulletin on swine feeding (Publication 1442) carried this obscure mention of a "potential" danger:

> More recently a problem has come to light of antibiotic-resistant strains, the result of antibiotic feeding. The consequences of this and the associated problem of transferred resistance are potentially serious and may ultimately force a revision of our thinking on antibiotic feeding. The fact remains, however, that a 10 percent or greater improvement in feed efficiency yields a very substantial increase in the swine producer's net income.

What they are talking about is the development of antibiotic-resistant strains of bacteria that cause infectious diseases, and the transferral of this resistance to strains that may attack humans. I don't believe 10 percent improvement in feed efficiency stacks up against these potential risks to pig and human health.

From Starter to Grower Rations

If you buy weaned 30- to 40-pound piglets, they probably will have been on a Starter ration. They may even have been on a Grower ration, since some people bypass the Starters altogether, preferring to go directly from Pre-starters to Growers. It is good to know what they have been getting, so that you can avoid compounding the stress of moving by drastically changing their diet as well.

If the weaner piglet is on a Starter ration, 50 pounds more will see him through to the Grower stage. Mix the last few pounds of Starter into the top of the first bag of Grower to smooth the transition, and you're off. In the time it takes for that piglet to go through about 500 pounds of Grower, you will have somewhere between 150 and 200 pounds of pork.

Thirty-pound weaners eat about two pounds of dry food daily and grow at a rate of about a pound a day. By the time those pigs reach market weights in the neighborhood of 200–250 pounds they will be eating six to eight pounds of dry feed a day and be gaining

Recommended Protein Levels, Average Rate of Gain, and Feed Consumption for Swine

Description of Animal	Protein Content of Ration (%)	Average Daily Gain (lb.)	(kg)	Average Daily Feed Intake (lb.)	(kg)	Total Feed Needed to Attain Weight[3] (lb.)	(kg)
BREEDER STOCK							
Gestation:							
Gilts	14	.60	.27	5.0	2.27		
Sows	14	.50	.23	5.0	2.27		
Lactation:							
Gilts and Sows	15			12.0	5.44		
Boars	14			6.0	2.72		

GROWER-FINISHING PIGS Pig Weight (lb.)	(kg)	Age, Days				Lb. or Kg Needed per Lb. or Kg Gain[2]		
Birth		0	—	—	—	—	110	50.0
10	4.5	14	17-20 (creep)	.50	.23	1.5	130	59.1
30	13.6	48	22 (early weaned)	1.00	.45	1.9	165	75.0
40	18.2	56	18	1.20	.55	2.2	185	84.1
80	36.4	85	14	1.65	.75	2.8	285	129.5
100	45.4	96	14	1.80	.82	3.1	345	156.8
120	54.5	108	14	1.90	.86	3.3	410	186.4
160	72.7	128	13	2.00	.91	3.5	545	247.7
200	90.9	145	13	2.10	.95	4.1	700	318.1
220	100.0	155	13	2.15	.98	4.4	785	356.8

[1]From Ill. Circ. 866, 1966, with permission of Dr. D.E. Becker; except for "protein content of ration," which the author adapted from NRC Pub. 1599, Rev. 1968.

[2]Based on feeding fortified corn-soybean rations. Where other feeds are used, quantity of feed needed per pound of gain will vary inversely with the calorie density of the diet fed.

[3]Includes feed for sows starting 30 days prior to mating and for breeding boars. Based on feeding fortified corn-soybean rations. Where other feeds are used, quantity of feed needed will vary inversely with the caloric density of the diet fed.

Source: Swine Science, 4th edition, 1970, by M.E. Ensminger. Published by The Interstate Printers and Publishers, Inc., Danville, Ill.

from 2 to 2¼ pounds of live weight. The average rate of conversion of good dry feed to pork is between three and four pounds to one. A commercial farmer will often aim for something better than three to one. As little as a tenth of a pound of improvement can mean quite a difference in profits when it is multiplied times the thousand or so pigs that may go through a feeder barn in a year.

Feed Consumed, Gain, and Efficiency at Various Weights

Live weight (Lbs.)	Air-dry feed (Lbs.)	Daily gain (Lbs.)	Feed per 100 lbs. gain (Lbs.)
25	2.0	0.8	250
50	3.2	1.2	267
100	5.3	1.6	331
150	6.8	1.8	378
200	7.5	1.8	417
250	8.3	1.8	461

From Morrison, *Feeds and Feeding, Abridged.*

When using commercial, pre-mixed rations, about the only questions that come up are what forms to use—loose ground or shaped into pellets or crumbles—and how to feed them—wet or dry; and on demand (also called *ad lib feeding* or *self-fed*) or on a controlled, hand-fed basis?

Loose ground feeds are cheaper than pellets or crumbles. They also lend themselves better to being mixed with water or milk for variations of moist or wet feeding systems. When they are fed dry, these ground feeds are dusty, there is more waste and they tend to gum up water bowls.

Pellets and crumbles are basically the same nutritionally as the ground, but are steamed and pressed into nuggets. They are generally more expensive than ground feeds, but because they are not dusty, because there is less waste, and because they are more palatable, the higher initial cost may be more than made up for in the end.

Feeding

WET VS. DRY FEEDING

It is in wet feeding that ground feeds prove best. Pigs do better on wet feeds, but they are not used on many commercial farms in North America, in part because we have not developed the systems for preparing and delivering feeds this way at anything like the low costs of delivering them dry. We may see an increase in wet feeding, though, as European technology is brought in.

I can imagine a small revolution in cheese manufacturing areas as whey-feeding to hogs—as a physical process—becomes simpler than dumping this cheesemaking by-product in the nearest river.

In wet feeding the dry ration can be pre-mixed to a loose porridge consistency in a bucket and then slopped into a trough, or water and dry food may be poured into the trough separately. This way the pigs do the mixing and it saves having to wash out mixing buckets. When mixing batches of wet solid foods for slopping several pens— grains or garbage—watch out that the solids don't settle out in the mixing tub. This may result in an uneven distribution of foods down the line.

There are three ways dry ground feeds may be wasted: They may accumulate in watering troughs or bowls, washed from the coated snouts of thirsty pigs, or they may be sneezed into the air or blown away in a breeze.

If I were dry-feeding pre-mixed commercial rations to lots of pigs indoors I would use formed feeds, for the health of the pigs and my own lungs. Dust can get very bad. If I had those lots of pigs outdoors

I might use ground dry feeds if they could be self-fed through hoppers leading to wind-proof troughs covered with little trap doors.

AD LIB VS. CONTROLLED FEEDING

Dry feeds are easily fed ad lib. Pigs often grow faster, and there is a lot less labor involved when they can help themselves to the trough beneath a hopper of feed whenever appetite spurs them on.

The efficiency of feed conversion by pigs on self-feeders can depend a great deal on the feeds used and on how they have been prepared. Pigs fed whole grains such as corn may convert the hard-cased seeds more efficiently when they are fed ad lib, in part because they take time at the trough to chew their foods better. Hand-fed pigs, hungry and competing, may wolf down whole grains, the result being a lower-than-normal rate of digestion.

Pelleted feeds may be too palatable for self-feeding some pigs a standard Grower all the way to slaughter weight. Commercial feeders not rearing pigs genetically selected to finish well on these rations fed ad lib may find their animals overeat and get too fat. From around 150 pounds these pigs must be held in check through hand-feeding or through the introduction into the hoppers of high-fiber (low-calorie) feeds. The pigs then feel full without overeating.

HAND FEEDING

Some may prefer to hand-feed pigs, dishing out their meals two or three times a day into a trough or onto a clean floor or platform. Advantages here are that the costs are low to nothing for feeding facilities, that you can restrict the pig's feed—maybe even regulate the amounts daily, depending on the amounts of household garbage or other supplements available—and that, by necessity, very important time is spent observing the pigs at least twice every day.

Despite the pig's magnificent ability to convert grain to meat, some contend that confined pigs raised solely on commercial rations do not taste as good as those raised on more varied diets that include pasture, green feeds and perhaps some dairy products.

However, there is so much else to learn when you are starting out with your first pig(s), maybe it is better to go with (and rely on) commercial rations one time through. Supplement with pasture, family swill, garden and home-dairy excesses, lawn clippings and whatever else edible comes to hand, but count on the commercial feeds to provide the basic diet sufficient for maintenance and growth.

In the end, you may discover that the cost of raising the pig was higher than necessary. For one thing, the full value of the supplements may not have been appreciated. And the carcass may be overly fat. (Chapter 17 suggests ways to use extra fat or lard.) But better flavor, and the experience gained, will more than offset any losses.

A couple of final notes: If you provide your pigs with self-feeding hoppers filled every once in a while, check daily to make sure that feed has not become clogged in the hoppers.

Supplemental foods should go in a separate trough so that regular dry foods are not interfered with. You may put in a supplement that a pig does not like, and if it goes on top of the foundation ration of commercial feeds you've messed up the day's program.

Pigs can be finicky about new foods. It may take more than one exposure to a supplement before they accept it. Or try priming their appetites with a few hours of starvation.

Notes on Feeding
Breeding Stock and Newborn Piglets

POTENTIAL BREEDERS

Young boars and gilts still growing should be taken off self-feeders by the time they reach 150 to 200 pounds and be hand-fed according to the recommendations of feeding standards (Appendix), or amounts necessary to keep them healthy but not fat. One rule of thumb for a maintenance ration (one that merely keeps an animal, providing nothing for growth, reproduction or lactation) is to provide the amount of a nutritionally balanced feed, having about 13 percent of protein, that a pig will clean up within a half hour, twice a day. Because growth in pigs continues to 18 months and to weights of 300 pounds and more, it will be necessary to balance a rule of thumb such as the one above with common sense. Again, feed for health but not for fat.

Morrison recommends that young breeding stock "be on good pasture during just as much of the year as possible. This is the best insurance against any lack of vitamins. When pasture is not available, 10 to 15 percent of legume hay should be included in the ration."

SOWS AND GILTS

Sows and gilts can be maintained on relatively small amounts of lower-protein feeds than are fed the fast-growing feeder pig. Four to six pounds daily of dry matter in a balanced ration giving 12 to 15 percent protein is a common rate. A pound or two more per day may be fed sows in poor condition.

These animals (and boars too) cannot be allowed a balanced, low-bulk ration free choice because they will get too fat. Fat sows have fewer piglets. Fewer eggs may be successfully shed from the ovaries, fewer may be fertilized, and fewer fetuses may be carried through the full period of gestation. Fat sows take longer to farrow,

**Sow Milk Production vs. Litter
Nutrient Requirements**

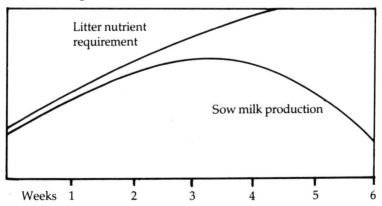

resulting in more piglets being born dead. So don't overfeed brood stock.

Some people "flush" their sows, meaning that they increase rates of feeding for a brief time just prior to breeding. The idea, common in husbandry with other species as well, is that an animal on a rising plane of nutrition is more apt to conceive. In sows this could mean an increased number of eggs shed and fertilized. Once breeding has been completed, the sow's level of feeding is dropped back to the pre-flushing rate. There are arguments for and against flushing—whether or not it is worth the effort. Certainly it is more difficult to manage under more intensive breeding programs where sows are bred back within a week of their previous litter being weaned.

Most of the growth of unborn piglets occurs in the last month of pregnancy. Feeding rates may be increased to six to seven pounds of dry matter in a complete ration at this time. Some managers may go as high as 10 pounds daily pre-farrowing with top-quality sows. However, others report increased problems with constipation and stress at farrowing time when sows are fed to excess late in gestation. Beginners should beware and feed conservatively.

Beginning a few days before a sow or gilt is due to farrow she will benefit from some wheat bran (or other laxative food) being added to her ration at a rate of about a pound each day. The day she farrows she may show little interest in food. Offer her fresh water anyway, and a bit of bran.

After farrowing, come back on a 15 percent protein feed slowly—including some laxative food the first couple of days—giving her but a third to half her regular amount that first day. Increase the feeding each day as the sow cleans up readily (within half an hour) what is fed. If a sow shows signs of constipation in spite of the bran, take her for a walk. A bit of exercise works wonders.

A large sow with eight or more piglets will need lots of food, as much as 12 or 14 pounds of dry matter daily, to keep milk production in line with demands. They are far greater demands than those placed on beef cows, because piglets grow faster than calves, and because sows' milk is richer than cows'. At peak production—which normally comes between the third and fourth week after farrowing—a sow may produce as much as two gallons of milk each day.

PIGLETS

Piglets will begin to show curiosity in solid foods within a week of birth. Sprinkle some commercial Pre-starter or a mixture of corn-flakes or rolled oats and dried skim milk powder (half and half) on the floor of the creep area. A tablespoon or two each day is plenty to start because at first the piglets will only root it about. Offer them a litte fresh food each day and they soon will be eating. Increase the amount offered as they clean it up.

There may be some loosening of the bowels as piglets branch away from mother's milk. It is not a cause for concern unless the pasty looseness becomes watery diarrhea.

BOARS

An active, mature boar should get six to eight pounds daily of dry matter in a diet giving him 12 to 16 percent protein. The amount depends on his age, size and use, and must be increased or decreased depending on condition. Overweight boars can be hard on sows during breeding and are easily stressed and exhausted. Light use for a junior boar—under twelve months—is having to breed less than two sows (four breedings) in a week's time. For a mature boar it is having to breed less than three (six matings).

A mature boar will be able to make use of grass, weeds and other fibrous foods, and will relish them. A boar on good pasture or being fed good forages or grass-legume hay to appetite, may require but a pound per hundredweight of supplemental grain rations daily to keep him in top shape.

CONDITION

"Condition" seems a meaningless term. What does it say to the person not familiar with the look of pigs in their prime? One of the obvious signs of underfeeding is the backbone rising out of the flesh, giving the pig a peaked-roof look. Other signs of malnutrition or low nutrition, may be a lack of sheen to the coat, curly hair, or a scurfy and scaly skin.

Older animals, especially sows that have had many litters, may become very hard looking, with lots of bony bumps, sags and hollows that don't fill out regardless of feeding.

Younger hogs that are overfed become jowly, and flabby in the hams.

It will take time and the experience of seeing lots and lots of pigs and sows and gilts and boars before a person new to the game becomes adept at judging condition and knowing how to feed to make the most of the stock.

HOME-MIXED RATIONS

For those who may be interested in mixing their own foods on the farm an introduction to the principles and methods behind the creation of balanced swine rations is presented in Appendix A.

CHAPTER 12

The Scavenger Pig

When you consider that more than half the world's population is underfed or starving it becomes incredible to unconscionable that we in North America pour mountains of edible grains down the gullets of our hogs and steers.

One answer is vegetarianism. Another is to create pig rations less dependent on things people eat. I have called it "scavenger pig." I think of it as pigs following people, living off foods we either do not or cannot use.

Raising scavenger pigs is not always easy. We have to put up with physical and mental laziness. Commercial, balanced rations are as close as our pocket books. How much simpler to feed these than to work so hard figuring out what ingredients are needed, and then going after them.

There may be a problem with space. Many hog farms today work from a land base of zero. They haven't the room to grow pig foods even if they wished to do so.

There is the question of time. Although many foodstuffs may be far cheaper to produce, purchase or procure than the usual commercial feeds, there will almost always be a larger commitment of time involved. Also it may take more time to grow your pork with scavenger foods.

Scavenger foods may be cheap, even free, to begin with, but after adding transportation and possible processing you may have a final cost that is comparable to the most expensive commercial rations.

There is a consideration of guaranteed supply, which is especially important for commercial producers. A person with a pig or two can take wonderful advantage of unexpected food bonanzas that come and go with such irregularity that the large producer cannot work them into necessary planning.

And last there are custom-developed tastes for cuts of pork of a certain size and of ratios of fat-to-lean, which in many cases will have to change if best uses will be made of scavenger foods. In 1953 Fishwick wrote:

> More reliance must now be placed on farm-grown foods for pig feeding. It can no longer be accepted that pigs must be given an all-meal diet. Root crops, grass and other farm by-products must now play an important part in pig feeding.
>
> Immediately a problem arises. The foods that can most easily be grown at home for the pigs are generally low in protein and are mostly rather bulky. This means that the older and heavier pigs are better able to handle the farm-grown diet, and to make maximum use of it, it would follow that the pigs should be marketed at a heavy weight. Indeed this did happen during the War, when quantity and not quality was the more important factor.

All of these considerations are more likely to be barriers for commercial pig farmers, on one hand, who must remain economically competive and turn out a product of consistent quality that meets certain prescribed standards. Small, home producers, on the other hand, are apt to have the flexibility that will allow them to turn their pigs into proper scavengers.

Pastures

Although the growing pig, with its small, simple stomach and high demands for nourishment, cannot take much food value out of grasses and other high-fiber forage crops, the foods and minerals it does get from a small piece of land will serve to offset any deficiencies in other foods provided.

"Pasture is the safety valve in swine feeding," wrote the authors of *Raising Swine*. "Much of its value is due to its contribution of known and unknown but essential nutrients."

In addition, Ensminger says, "good" pasture "can reduce the grain required to produce a hundred pounds of pork by 15 to 20 percent, and the protein supplement required by 20 to 50 percent."

Nutritionally, the best pasture will provide a mixture of actively growing grasses and legumes, and less area of a pasture like that would be needed for grazing a number of pigs than a pasture of bushes and weeds. The same good pasture will provide more and better food in spring than it will later on in the hot, dry times of late summer.

Fishwick mentions an experiment in England in which two groups of pigs, all 52 pounds at the start, were put through a 100-day feeding trial with one group given pasture and the other an equal area of plowed ground. All were fed the same ration, beginning at 1½ pounds per day and going to 4 pounds.

At the end of the trial the pigs on pasture weighed an average of 169 pounds, and those on dirt, 139. The meal-eaten-to-live-weight-gain for those on pasture was 2.9 to 1. For those on dirt it was 3.8 to 1.

PASTURE VALUES

Because the quality varies so much, it is impossible to say precisely how much pasture a pig needs, but it is safe to figure that through a summer a feeder pig could make good use of one twentieth to one tenth of an acre of grazing ground (roughly 2,500 to 4,400 square feet).

General Recommendations for Planting and Grazing Pasture Crops for Swine[a]

Pasture crop	Time to seed[b]	Seed required per acre, lb.	Approx. time from seeding to grazing, months	Approx. carrying capacity per acre. No. of 50- to 100- lb. pigs
Bluegrass[c]	Fall	6- 10	12-18	5-15
Alfalfa	Spring or fall	15- 20	4-5	15-25
Red clover	Spring or fall	10- 15	4	12-20
Sweet clover	Early Spring	20- 30	3-4	15-20
Ladino clover	Spring or early fall	2- 4	6	20-25
Rape	March to June	4- 8	1½-2	15-25
Soybeans[d]	May to July	60-120	1½-2	12-15
Cowpeas	May to July	60-180	1½-2	12-15
Sudan grass[d]	May to July	20- 30	1-1½	15-20
Rye	Spring or fall	85-170	2-3	15-20
Barley	Spring or fall	75-100	2-3	10-12
Wheat	Spring or fall	90-120	2-3	10-15
Oats	Spring or fall	65-130	2-3	10-15
Italian rye grass[e]	Spring or fall	10- 25	2-3	10-15
Oats and field peas	March to May	50 oats 60 peas	1½-2	10-15
Mixed-grass pastures:				
No. 1:	Spring or early fall			
Orchard grass		6-8 ⎫		
Red clover		3-4 ⎬	5-6	20-25
Ladino clover		1 ⎭		
No. 2:	Spring or early fall			
Bromegrass		10 ⎫		
Alfalfa		10-15 ⎬	5-6	20-25

[a]Grass, *U.S. Department of Agriculture Yearbook*, 1948, p. 819.
[b]For information as to the best forages for locality, as well as time and rate of seeding, consult experiment station or local county agent.
[c]Usually sown in grass mixtures for hay. Grazed second or third year.
[d]Often seeded together at rate of 10 lb. Sudan grass and 1 bu. soybean per acre.
[e]Frequently seeded with grain crops or give solid turf. Also furnishes succulent grazing after grain crops mature.

Reprinted from Deyoe and Kreider, *Raising Swine.*

Numbers of pigs may require less space overall than one or two animals, since a certain more nearly fixed amount of room will be needed for hutches, waterers, supplemental feeding stations and so on.

For growing pigs good pasture may be worth one to three pounds of a complete meal per day. The upper range will be with larger pigs and the best grades of lush, actively growing mixed (legume/grass) pastures.

Experimentally, mature, dry sows have been maintained solely on good pastures, with supplemental grains being added to the ration beginning the last two weeks prior to farrowing. In practice, up to ten sows may be run on an acre of pasture that provides them with about half their feed.

An hour of grazing is supposed to give a pig an equivalent of four to six pounds of green food, and this (based on other reports I have read relating to values of green feeds like cabbages, kale or lawn clippings) should be worth almost a pound of a 16 percent crude protein ration.

A sow's capacity for young lawn clippings has been found to be about 24 pounds daily, and this is figured worth about three pounds of a 16 percent protein ration. Since a mature animal needs four to six pounds daily of a complete meal, it would appear this statement of lawn clipping value conflicts with claims that sows can be maintained on pasture alone. However, I think it only emphasizes how much the value of green feeds can vary. You can't simply put a bunch of pigs or sows out in a big field and say, "there, that's taken care of." Their condition and that of the pasture have to be watched, and supplemental feeds must be brought on as conditions indicate.

ROOTING

There is no guaranteeing pigs will not root up a pasture. However, pasture conditions and, to an extent, the breed of pig, can make a difference in degree of destruction of a piece of ground you don't want turned.

Most important in keeping your pigs from going underground—unless you have put rings in their noses—is to keep them adequately fed and on cover that is well established and not overgrazed. They should be taken off a pasture during heavy rains or wet spells when the sod starts going to pieces beneath their feet. They should not be so confined or crowded that they begin digging out of sheer boredom.

Don't be fooled into thinking you have come across "good" pigs that don't root just because they spend a couple of days on a pasture without submerging. The third day could see a complete change of behavior as, perhaps, one pig follows a night crawler down its burrow.

It is, I think, the hunt for worms and such that does first lead pigs to root where the grass is the most lush and green. Here is the most fertility—where the action will be—from molds and microbes on up the food chain. And so the suggestion is made by a friend who tried: Don't be too quick to put pigs on a spotty pasture hoping they will graze the best and root up the worst. Likely they will ignore the rough, poor cover and dig up the good parts.

CLEARING LAND WITH PIGS

Far from discouraging rooting, many people do everything they can to encourage the practice in the course of using pigs to clear and plow land. A thorough job can be done on thicketed scrub land or on old fields that are returning to second growth.

The pigs should be confined so that they will root the enclosed area completely. Moveable electric fences are ideal on the roughest, brushiest and rockiest land. On more open ground portable huts with attached exercise yards may be better.

After two or three weeks pigs will have all but the largest shrubs uprooted. Rocks will be tossed on end where they're easy to roll onto a stoneboat. The larger trees may be unearthed as well, if you "plant" cracked corn, apples or other goodies in crowbar holes poked amongst their roots.

A combination of hogs and sheep may be used to establish pasture on shrubby barrens or reverted crop lands. I've not tried it, but the system sounds reasonable and is sworn to by the person who worked it out in New Brunswick.

He first ran hogs on the barrens (rough, sour and rocky land covered with waist-high shrubs). Once most of the vegetation was dead or well beaten down, lime was spread casually with a shovel. It did not matter if the lime was not well scattered. The pigs finished the spreading.

When winter came and snow covered the area, bales of hay were broken and fed to sheep over the same ground. Hay seed, direct from the bales, or passed through the sheep, sprouted over the rich bed, found once the snows melted in spring.

DANGERS WITH PASTURING

Certain weeds and trash such as clay pigeons and tarpaper or other materials containing asphalt are poisonous to hogs. Cocklebur, members of the Nightshade family, water hemlock, buttercup, tarweed and St. John's wart are in the list of plants poisonous to pigs (*Diseases of Swine*, Iowa State University Press, 1970).

Inexpensive Grains and Mill Wastes

Low-cost grains and milling wastes or by-products, such as sweepings, may make good foundations for swine rations. However, watch out for unacceptably high fiber content (especially for growing pigs); for chemically treated seed grains, and for certain molds that can cause serious diseases and/or reproductive problems.

A University of Guelph (Ontario) researcher says corn containing as little as 2 to 5 percent of moldy kernels can be toxic to pigs if it is a mold that contains mycotoxins. "The molds of most concern appear to be pink or white," J.C. Sutton reported in 1977. "Green and black molds do not normally produce mycotoxins."

Most agricultural extension services provide free laboratory tests that will tell you about fiber content or poisons in feeds.

**Value of 100 Lb. of Various Feeds in Comparison
with Shelled Yellow-Dent Corn (No. 2) When the Latter
Is Taken as a Standard of 100 Percent[a]**

Feed	Approx. value, %	Comments
Yellow ear corn	100	1 bu. (70 lb.) ear corn = 1 bu. (56 lb.) shelled corn
Flint corn, shelled	97	May be ground for hogs over 150 lb.

Value of 100 Lb. of Various Feeds (Cont'd)

Feed	Approx. value, %	Comments
Waxy corn, shelled	100	
Ground shelled corn	104-105	If hand-fed to pigs over 150 lb., 100 lb. ground corn is worth 106-107 lb. shelled corn.
Ground ear corn		Not recommended
Soft corn	Variable	The dry matter in soft corn has the same value as dry matter in sound corn.
White shelled corn	100	White corn is deficient in provitamin A.
Moldy shelled corn	68-89	Value varies with damage
Oats, ground finely	80-100	Fed as 20-33% of ration
Oats, whole	73	Fed as 20-33% of ration.
Oats, hulled	140	Hulling is usually less profitable than grinding.
Barley, ground or rolled	91	Should always be ground or rolled; scabby barley is unpalatable and even dangerous.
Wheat, whole	100-103	May be self-fed whole
Wheat, ground	109-110	If hand-fed, grinding saves as much as 23% of the grain.
Rye, ground	91-92	It should always be ground. Poor-quality rye is worth less than 90%. Do not feed rye containing ergot to sows or young pigs.
Grain sorghums, ground	90-94	
Buckwheat, ground as oats		Use as oats are used in rations.
Hominy feed	100	Replace only 50% of corn in fattening rations.
Molasses	100	Fed as 10% or less in ration.
Bakery waste (stale bread etc.)	75-80	Sometimes causes constipation.
Potatoes, cooked	28	Cook, salt, and discard water before feeding. Feed 3 parts potatoes to 1 part of other grain.

[a]If further details are desired, refer to W.E. Carroll and J.L. Krider, *Swine Production*, Mcgraw-Hill Book Company, Inc., New York, 1950.

Reprinted from Deyoe and Krider, *Raising Swine*.

Garbage

I have come across a wide range of statistics on the amounts of edible garbage we North Americans throw away each year—all the way from a low of about 40 pounds per capita to a high of 154. The value of this food for pigs ranges as widely, depending on its nature and on the age of the pigs being fed. It could take 10 to 20 pounds of leftover vegetables from a produce stand to equal the value of a pound of mixed grains. Household garbage could be on a par with or worth even more, pound-for-pound, than grains.

The table on garbage composition shows a wide difference between wastes from hotels and restaurants, institutions, military bases, and your average town dump. From what Barth discovered, consider yourself lucky if you can get garbage from an army kitchen—but be ready with the ladle to dip off the extra fat.

Youngest, fastest-growing piglets are least able to make good use

**Average Composition
of Garbages by Type[1]**

Criteria	Hotel & Restrant	Institnl	Military	Muncipl	P .05	P .01
		TYPE OF GARBAGE			Difference required for significance	
No. of samples	30	58	56	21	—	—
Dry Matter, %	16.0	17.5	25.6	16.6	2.9	3.7
Crude protein,[2]%	15.3	14.6	15.9	17.5	2.1	2.7
Ether extract,[2]%24.9	14.7	32.0	21.4	4.6	5.9	
Crude fiber,[2]% 3.3	2.8	2.8	8.4	1.3	1.6	
Ash,[2]% 5.7	5.2	5.5	8.6	1.1	1.4	
NFE,[2]%	50.7	62.6	43.8	44.0	4.8	6.2
Gross energy,[2]kcal./g.	5.33	4.81	5.67	5.10	0.26	0.33

[1]Taken from Kornegay et al., 1965.
[2]Dry matter basis.

From Bul. #829 "Nutritive Evaluation of Garbage as a Feed for Swine," New Jersey Agricultural Experiment Stations, Rutgers U., New Brunswick, N.J.

of most garbage because of its bulk. Fishwick suggests pigs be 60 pounds or bigger before being fed any quantities of "swill."

GRADING SYSTEM

During World War II, Fishwick and others in England worked out a feeding system based on a division of garbage into three grades. *Grade A* is almost entirely roughages—green vegetables, vegetable peelings, potatoes, carrots and such. *Grade B* is two-thirds or more roughages and about one-third bakery wastes or other starchy or sugary wastes. *Grade C* is the richest in food value, being half high starch or protein wastes and the rest of roughages having a low fiber content.

Pigs would not make it on Grade A swill alone. From 60 pounds to market weight Fishwick recommends feeding all the Grade A swill they will take, plus two to four pounds daily (the more as they grow) of a 16 to 17 percent crude protein feed. This could be a commercially pre-mixed feed or a combination of, say, 85 percent corn or barley meal and 15 percent fish meal.

Pigs 60 pounds and more can make it on Grade B swill alone but Fishwick suggests best results will come with feeding a couple of pounds daily of a 14 to 16 percent protein feed along with all of the garbage they will eat.

Grade C swill may be the only food needed for 60-pound piglets and larger if the animals are on pasture. If they are not, then from about 150 pounds on they should receive "an occasional allowance of green food."

In all cases Fishwick's swill is watered down before cooking "just sufficient to insure that all the constituents are floating."

COOKING GARBAGE

All garbage should be cooked. Not just because it is required by law but because it is smart. Boiling garbage for at least half an hour assures that parasites—especially the trichina worms you don't want burrowing through the muscles of your arm—are killed. You want more than a surface bubble. The whole mix should come to 212°F. (100°C.). Excess fat that rises during cooking should be removed. Use it to make soap.

Cooking also gives a food that pigs clean up readily and entirely. With raw, household garbage I have noticed pigs pick and choose. Garbage presented as a stew is quickly devoured.

A good suggestion from Steven Thomas is that in arranging for restaurant, store or other garbage, you offer to provide containers. If you would offer a penny a pound for edible restaurant garbage and supply heavy plastic bags and a couple of barrels, everyone should come away happy. Of course you would expect that dishwashers are taking the trouble to sort out paper, broken dishes and such. Avoid soap, soda and other cleaning materials, says Mr. Fishwick— also salt, quantities of tea leaves and tobacco. Rhubarb leaves are highly toxic.

There may be such a thing as being too particular, though. I have heard that some of the New Jersey piggeries that fed garbage made a good sideline out of recovered silverware.

Roots

Many root crops, including white and sweet potatoes, carrots, turnips and Jerusalem artichokes, make excellent pig foods.

A German feed scientist, G. Lehmann, in the early part of this century worked out a sysem for feeding roots. It is a logical system, easy to follow, and is based on two principles: (1) that a baby pig, for lack of gut capacity, must be started on concentrated foods, and (2) that, as a pig grows, its protein needs remain more or less constant.

Potato steamer.

Raw potatoes: 7 lbs. equals 1 lb. grain.
Cooked potatoes: 4 lbs. equals 1 lb. grain.

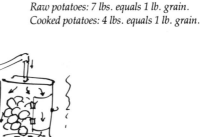

Antique potato or root cleaner.

Under the Lehmann system the weanling eats a ration of grains and protein concentrates (2½ to 3½ pounds daily depending on bulkiness of roots being used) having an overall protein value of about 18 percent.

The roots, usually potatoes, are gradually worked into the diet and fed to appetite as a supplement to the daily, constant allowance of 18 percent meal. Automatically the Total Digestible Nutrient (TDN), or energy value (see Appendix A), of the overall ration is increased while the overall percentage of protein drops in the same way it would if you were feeding a sequence of commercial Pre-starters, Starters and Growers.

More recent experiments in England seemed to indicate that Lehmann was too liberal with the protein, and that a "basal meal" having only 15 or 16 percent protein was sufficient.

There are some cautions with the Lehmann system. Make sure each pig gets its share of the "basal meal" and so its allowance of protein each day. Also, the fiber of the basal meal should be adjusted depending on the fiber content of the roots or other crop being fed to appetite, so that overall the amount of fiber in the diet does not become too high (generally less than 10 percent for growing pigs). Also, make allowances for the fact that most roots are deficient in vitamins and minerals.

WAYS TO FEED HOME-GROWN CROPS

There are different ways to get home-grown feeds into your pigs. They may be harvested and hauled to the animals, or the pigs may be allowed to do the harvesting themselves. "Hogging down" is the term used whenever pigs are sent into a field of peanuts, corn or the like. Sometimes pigs are turned onto already harvested fields to let them clean up left-over small potatoes, peanuts, etc., which they will do with wonderful thoroughness.

Masting

Tree nuts and seeds are called *mast,* and the running of pigs over a forest floor to feed on the mast in the fall of the year is an old, old way of growing pork. *Foxfire* recounts the ways of "masting" in the United States southern mountains:

> Our contacts tell us that the sweetest meat came from hogs fattened on chestnuts. One problem was that instead of rendering into good, white lard, the fat of these hogs would boil down into a dark oil. Acorn mast made the meat taste bitter and altered the consistency of the fat. For these reasons hogs to be slaughtered were often rounded up and brought down out of the mountains to the farms where they were fed on corn for anywhere from a few weeks to over a month. This removed any bitterness from the meat and softened the fat properly for rendering into lard.

I wonder about this last statement—about a corn diet softening the lard, because other references to masting mention a problem of soft pork. "Nuts should never be given to swine," says Youatt, as this makes a soft, greasy fat, "and imparts a sweet, unpleasant flavour to the flesh." Beech mast alone, he wrote, makes the fat oily "and impoverishes the lean." He also wrote that acorns roasted, boiled or steeped and mashed were better than acorns raw.

Hogs Following Cattle

When cattle are being fed grains, particularly whole grains, hogs may be fattened on gleanings from the cattle manure that will be laced with undigested kernels. Cows don't chew thoroughly, and up to 25 percent of whole grains eaten may pass through their systems unscathed.

An early edition of Henry and Morrison's *Feeds and Feeding* says pigs 50 to 150 pounds are best suited for "following." They recommended two or three pigs per steer that is being fed "snapped" corn (whole ears in the husks); one and a half pigs per steer on husked corn; one per steer on shelled corn, and one pig for every two or three steers on crushed or ground corn.

They also recommended pigs be given supplementary feeds in an area separated from the steers, and that they be fed their supplements when the steers are fed so that the grunters don't climb underfoot of the beef animals.

A variation on this theme of pigs following cattle was mentioned briefly in the chapter on Housing, talking about keeping pigs in manure pits beside or under cow barns.

Fish and Fish By-products

Mention fish for pigs and people turn up their noses, convinced the pork would taste like old cod. It may have a fishy taste if the pigs are fed quantities of fish—especially oily fish such as herring or mackeral—right up to the time of slaughter. But, according to several people I have talked to, many tons of pork have been raised on the coast of Newfoundland with fish as the only food, but for some household wastes, from 50 pounds to within two weeks to a month of slaughter. Traditionally the pigs are then switched to a diet of barley, about a hundred pounds for each one, "to change the fat," to avoid any chance of the pork tasting fishy.

A condition called *yellow fat disease* observed in pigs fed fish only, reputedly caused by a lack of vitamin E and resulting in flesh discoloration and heart and circulatory failure (*Diseases of Swine*, H. W. Dunne, 1970), apparently was not encountered by these people.

One Newfoundland fisherman/farmer told me that he starts his piglets on a commercial Starter and then a Grower until they are about 10 weeks old, at which time he introduces cooked fish heads or whole fish. The pigs may be on an all-fish diet for three to four months. Then they are switched to a commercial Grower, ground barley, oats or other grains for a period of about three weeks to "clean them out." He serves his fish with fresh water and some salt—maybe a teaspoon a day. As the pigs grow, they will eat up to two or three gallons of fish heads a day or an equal weight in whole fish.

A drawback with fresh fish is that it spoils fast. Fish silage or Liquid Fish Protein (LFP) may be answers. Raw fish or fish parts are mixed with water and acid in plastic barrels, fiberglass tanks or wooden puncheons. The fish, bones and all, break down into an acid soup that will last indefinitely as long as the pH is held at or below 2.5.

In some trials conducted at Agriculture Canada's experimental farm in Nappan, N.S. in the 1950s, fish silage made with sulfuric acid was fed to pigs all the way to slaughter, and taste panelists said there were "off flavours." They ran more experiments, each time withdrawing the fish from the pigs' diets earlier to see if taste would improve. There was improvement but no unqualified success. One official at Nappan today suggests "the real problem with taste panels is that off flavours occur in pig meat for no apparent reason."

A nutritionist with the Prince Edward Island Department of Agriculture working with LFP today suggests that fishy flavors are likely when mineral acids are used. A. Hamid Javed says organic acids (formic or propionic or a combination of the two acids) mixed with low-oil ("white") fish will produce a protein supplement less likely to impart flavors.

For more information on this unfolding possibility for people who can grow or get hold of cheap fish, write Agriculture Canada Experimental Farm, Nappan, Nova Scotia, and ask for information about fish silage. Or write to the P.E.I. Department of Agriculture, Charlottetown, and ask about their work with liquid fish protein.

Milk and Milk Products

Milk and its products make excellent pig foods, their only drawback in most cases being cost. Whole milk, skim milk, and real (not cultured) buttermilk contain 3 to 4 percent of top-quality proteins. Whey, the serum that separates from the curd in cheese-making, is highest in sugars, but even its small (less than 1 percent) amount of protein is of excellent quality.

Fishwick writes that in cheesemaking countries whey is "fed to appetite" and that pigs fed this way will drink 5½ to 6½ gallons (imp.) a day in addition to eating 1½ to 2 pounds of a complete meal. He says that the practice is to cut down on the whey allowance toward the end of the fattening period or else the pigs become pot-bellied.

Ensminger says 12 pounds of whey can be used to replace one pound of a complete feed.

Based on figures in *Feeds and Feeding, Abridged* (9th ed.) it seems that growing pigs make the best (most economical) use of between one and two gallons of skim milk or buttermilk daily, fed with one or a combination of grains. As with the Lehmann system, the amount of milk remains constant as the pigs grow because their requirements are proportionately less with age.

FEEDING SKIM MILK*

"A simple method of feeding separated (skim) milk to growing pigs is to allow each pig ¾ gal. (imp.) daily to slaughter, plus as much meal twice a day as it will readily clear up in about 15 minutes per feed. For pigs destined for pork or bacon it will normally be necessary to restrict the daily meal consumption once this has reached a level of about four to five pounds per head, depending on the type of pig."

*From British Ministry of Agriculture Leaflet 439, 1973, "Feeding Separated Milk." This may also be buttermilk. The Imperial gallon contains about five U.S. quarts.

On a replacement basis, Ensminger says six pounds of skim milk or buttermilk are worth a pound of a complete feed.

Skim milk, buttermilk and whey are all deficient in vitamins A and D. If quantities are fed to pigs not on pasture, other sources of these vitamins must be provided.

Minerals and Vitamins

Known mineral and vitamin requirements almost always are included in commercial swine rations. They can be incorporated in your own feeding program by offering boxes of minerals, where the pigs help themselves. Consult your veterinarian or agriculture extension office about possible trace mineral deficiencies specific to your area.

Piglets may be supplied both vitamins A and D by feeding them a teaspoon of cod liver oil daily. Older slaughter pigs not on pasture should get their A and D from a commercial preparation or through self-fed, sun-cured alfalfa hay, as fish oils might impart an off-flavor.

Suggested Mineral Supplement Mixtures
Which May Be Offered Free-Choice

	Suggested mixtures		
Mineral Sources	*1*	*2*	*3*
Ground limestone	40%	50%	40%
Steamed bone meal or deflourinated phosphate	40%	20%	
Wood ashes			40%
Stabilized iodized salt	20%	30%	20%

Legumes, green or cured, also provide the B vitamins except vitamin B_{12}, which may be provided in foods of animal origin, including tankage, milk and dairy by-products, worms, insects and ruminant manures.

Vitamin C is synthesized by the pig. Vitamin E is found in many foods but often is destroyed in harvest and storage. Therefore, pigs confined to home-mixed diets of prepared foods may benefit from an inclusion of this vitamin. Veterinarians and feed supply houses have vitamin E supplements and often it is included with A and D supplements.

CHAPTER 13

Rearing Your Own Pigs

Out of the many who raise piglets for pork will sprout a few who
decide to get into the pig business. Who knows why? It could be that
an able gilt you have grown for food has become too gorgeous to kill.
Maybe someone has a boar for sale cheap. These are terrible reasons
for trying to jump what may not at first appear a great chasm. But
it is.

A Risky Business

Keeping breeding stock and raising litters of piglets is costly and
risky. Food alone may run to $150 or more a year per animal.
Housing that is much more secure, and therefore more costly, is
needed for keeping fully-grown animals.

It takes several sows to justify the cost of a boar. The figure varies,
but certainly a half-dozen sows has got to be the barest minimum.
Some people spread the cost of a boar by putting him up for stud
service. This looks good but risks bringing in all sorts of diseases.

Two or three small farms might share the costs and responsibili -
ties of a boar. This more restricted situation would be healthier but
still not as good as keeping several sows and a good boar isolated
from the germy world out there.

HEAT PERIODS

From about five months of age on, a female pig comes into heat
approximately every 21 days. A heat lasts one to three days. With

gilts you will usually notice one or more heats before she is old enough for motherhood. Eight months of age, or third heat, is good for first breeding so that she will be close to a year old before her first farrowing. Note the first heat, which should come at 5 to 5½ months, and check a calendar for her next expected heat.

Once a gilt or sow is in pig she will not normally cycle again until three to five days following the weaning of her litter. A sow may be bred back immediately on this first heat. Some research has indicated that sows whose litters are weaned at three to four weeks of age may have larger subsequent litters if they are not bred back until their second heat following weaning. When weaning is not until the fifth or sixth week sows may be bred back that first heat without affecting litter size.

HEAT DETECTION

Signs of heat are more easily detected in gilts than older sows, and more easily detected among groups of females where they may be seen mounting one another. It helps to have the boar penned somewhere near and within sight and hearing of the sows who may then respond when in heat by pacing back and forth as close to his nibs as fences will allow.

Following are some of the more common signs of heat:

1. Flushed and swollen vulva—usually at its most pronounced about a day prior to the onset of the *standing heat* (see #4). Gilts may "show" in this way as many as two or three days prior to their standing heat.

2. Nervousness and more than the usual amount of agitated grunting.

3. A tacky vaginal discharge that may be more readily felt than seen.

4. A sow or gilt in *standing heat* stands stock still when you press down on her back. Bear down heavily with both hands. (Some sows used to attention may stand only because they expect a

Sow Heat and Fertility Periods

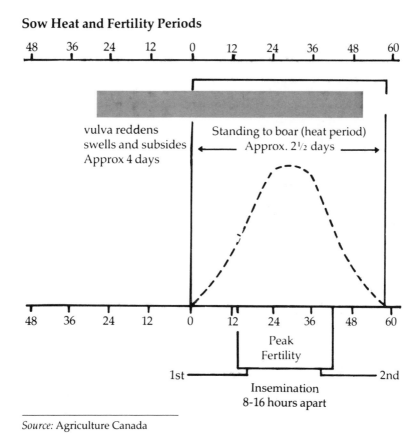

Source: Agriculture Canada

good scratching is in the offing. With a sow like this, try sitting on her back.)

5. Mounting among groups of sows. As sows come into heat they will be mounted by others, but will skitter away until they are in good standing heat, at which point they not only will stand still while being ridden but will start mounting their friends.

There will be no noticeable bleeding following a heat period, as may be seen in cattle.

Sometimes a sow that does not appear to be cycling will quickly come into heat after she is moved or otherwise radically jostled out

A sow or gilt in strong "standing heat" will not usually move when a handler bears down on her back.

of her normal routine. Boar scent and even recordings of a boar's *chant de coeur* sometimes are used to encourage heat response.

One way to catch sows in heat is to let them run as a group with the boar. But though this saves the herder having to detect heats, it is also sloppy; often it is impossible to know which sow was bred or when. When a boar mounts a sow he often leaves marks from his front trotters raking down the female's sides and shoulders—but maybe not. Maybe a sow comes quietly into heat and is bred in the night by an easy-going boar with clean feet. Is she or isn't she? Precious weeks or even months could be lost before discovering that a sow is not in pig, and is failing to cycle for some other reason.

A boar's ad lib service to a group of sows will likely be spotty. He may breed one sow more often than necessary and neglect another. His efforts can better be spent through a controlled, hand-mating system.

HAND-MATING

Under this system the sows are kept individually or in groups of up to ten or so. As they come into heat they are normally taken to the boar's pen, though in some instances, where sows are in separate pens, the boar visits her. If the timing is right they will breed within minutes following typical foreplay, which to us looks like little more than the boar nosing the sow into position. If the sow is not ready she will not want the boar to mount. He will chase and attempt to line her up. She will squeal and run, and maybe hurt herself trying to escape his attentions. Open the door and let her get back with her buddies. In a matter of hours her attitude may change completely as hormones bring on the standing heat that reflects a desire to breed.

A mature sow that is not ready to breed may turn on a young, inexperienced boar and give him such a sound beating that he will be afraid and inhibited for some time to come.

The young boar may need some assistance in lining up his mate, and an experienced herder will step in even to the point of making sure the boar's penis enters the vagina. In fact, this can be a problem—boars may fail in their aim. At times like this it may be possible

Being on hand to provide assistance and support greatly increases the chances for successful breeding.

to redirect the penis by hand. Otherwise it will be necessary to back the boar off the sow and start over again.

Pigs breed slowly. The boar may be a minute or more reaching the point of ejaculation, but it isn't over with any bang. Ejaculation goes on and on, for five minutes or more, involving the transport of a cup or more of fluids and billions of sperm.

This is one reason why great hopes are resting in the further development of artificial insemination in hogs. It is already done, and the processes are fairly simple, with boars trained to mount dummy sows for the collection of semen and sows being impregnated through artificial though accurately replicated boars' penises, complete with corkscrew tips. While AI in cattle is provided by professional technicians, it is likely that artificial insemination of swine will be a farmer's chore. Timing is more critical in swine and, as well, it is desireable that a sow be bred twice in a heat period.

It is important to have a floor that provides slip-free footing for the breeding pigs. Deep bedding or soft earth are best. If the pigs are on a hard surface and their feet seem to be sliding out from under them, add dry bedding, sand, or anything else that may improve traction. You may have to step in and provide support with your own feet against the slipping trotters. In providing such assistance, there is little to fear from the boar that is intent on breeding a sow.

The best boar won't last very long. Boars begin to taper off in potency and zeal at about five to six years of age. Also, they can become too large for the service of small sows that are physically incapable of supporting the weight of quarter-ton boars. A breeding crate that confines the sow and takes the bulk of the boar's weight may solve this problem, but older boars don't always accept the idea of a crate.

Breeding crate.

A sow should be bred twice during a heat at approximately twelve-hour intervals. This second breeding not only increases the chance of impregnation but even adds to the number of piglets that may be conceived. Some say to breed sows "when they come into heat and twelve hours later." Others say breed them "late on the first day you notice a sow in heat and again the following morning." For the novice, try breeding when you think the sow is ready, morning or evening, and again twelve hours later. Repeat attempts until two apparently successful matings have been achieved.

Some farmers run a second boar with groups of hand-bred sows as a back-up or guarantee of breeding.

Go easy with a sow or gilt just bred. It will be two weeks before the fertilized eggs will plant themselves on the wall of the uterus. Pummeling, sharp falls or even the shock of worming during this critical time may lead to shedding of the unanchored eggs.

GESTATION

The gestation period for hogs is 110 to 120 days, with 114 being average. The old calculated guess of three months, three weeks and three days comes very close.

Farrowing

A progression of signs foretells the day of farrowing. There may be a noticeable filling out of the sow in her third month of pregnancy. The udder begins to develop and fill seven to ten days prior to farrowing. Some time within a day of the event you can produce milk by gently squeezing a nipple between thumb and forefinger.

The last day of gestation the sow may attempt to build a nest. Even if she is in a crate she may try to scrape bedding back with her forefeet.

As the moment for farrowing nears, usually at night, a sow will become increasingly nervous. You may be on hand, but strangers should be kept away unless you know that the sow is easy-going. Many farmers insist on being on hand at farrowing time, even if it

Swine Gestation Table-115 Days

Bred / Farrow	1	2	3	4	5	6	7	8	9	10	11	12	13	14	15	16	17	18	19	20	21	22	23	24	25	26	27	28	29	30	31	
Bred Jan. / Farrow April	26	27	28	29	30	1	2	3	4	5	6	7	8	9	10	11	12	13	14	15	16	17	18	19	20	21	22	23	24	25	26	Jan. / May
Bred Feb. / Farrow May	27	28	29	30	31	1	2	3	4	5	6	7	8	9	10	11	12	13	14	15	16	17	18	19	20	21	22	23	---	---	---	Feb. / June
Bred Mar. / Farrow June	24	25	26	27	28	29	30	1	2	3	4	5	6	7	8	9	10	11	12	13	14	15	16	17	18	19	20	21	22	23	24	Mar. / July
Bred April / Farrow July	25	26	27	28	29	30	31	1	2	3	4	5	6	7	8	9	10	11	12	13	14	15	16	17	18	19	20	21	22	23	---	April / Aug.
Bred May / Farrow Aug.	24	25	26	27	28	29	30	31	1	2	3	4	5	6	7	8	9	10	11	12	13	14	15	16	17	18	19	20	21	22	23	May / Sept.
Bred June / Farrow Sept.	24	25	26	27	28	29	30	1	2	3	4	5	6	7	8	9	10	11	12	13	14	15	16	17	18	19	20	21	22	23	---	June / Oct.
Bred July / Farrow Oct.	24	25	26	27	28	29	30	31	1	2	3	4	5	6	7	8	9	10	11	12	13	14	15	16	17	18	19	20	21	22	23	July / Nov.
Bred Aug. / Farrow Nov.	24	25	26	27	28	29	30	1	2	3	4	5	6	7	8	9	10	11	12	13	14	15	16	17	18	19	20	21	22	23	24	Aug. / Dec.
Bred Sept. / Farrow Dec.	25	26	27	28	29	30	31	1	2	3	4	5	6	7	8	9	10	11	12	13	14	15	16	17	18	19	20	21	22	23	---	Sept. / Jan.
Bred Oct. / Farrow Jan.	24	25	26	27	28	29	30	31	1	2	3	4	5	6	7	8	9	10	11	12	13	14	15	16	17	18	19	20	21	22	23	Oct. / Feb.
Bred Nov. / Farrow Feb.	24	25	26	27	28	1	2	3	4	5	6	7	8	9	10	11	12	13	14	15	16	17	18	19	20	21	22	23	24	25	---	Nov. / Mar.
Bred Dec. / Farrow Mar.	26	27	28	29	30	31	1	2	3	4	5	6	7	8	9	10	11	12	13	14	15	16	17	18	19	20	21	22	23	24	25	Dec. / April

Each box lists a breeding date, followed by an expected farrow date 115 days later.

Source: Alabama Cooperative Extension Service, Auburn University, in cooperation with the U.S. Department of Agriculture.

means staying up all night. No doubt they save piglets. Other commercial operators with 30 or more sows feel they can't do it on a regular basis, though it can be argued that the saving of as little as one piglet per litter would pay for occasional part-time help.

The sow about to pig grunts with increasing frequency and lets out an occasional squeal. She will get up and down, up and down. If she were a cow she might crane her neck to try to discover what is coming off back there. But imagine being a pig and being unable to bend around. Something is happening, but what? It is just too overwhelming, and so she lies down for the umteenth time, and waits.

Piglets should begin arriving within an hour of the first signs of labor strain and within minutes of the appearance of any quantity of fluids, indicating a placental pouch has broken. Each piglet will be in its own pouch. Normally, they are born at 10- to 15-minute intervals, and will come out back first or front first—it makes no difference. The sow may give birth while lying on her belly, on one side or the other, or even while standing.

All piglets should be born within about three hours of the arrival of number one. Then comes the afterbirth. I have heard of instances

Piglets may be born head or tail first.

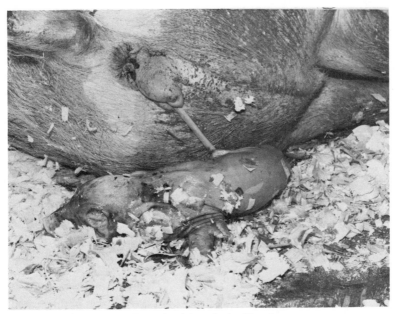

Each is covered by a thin membrane that dries and rubs off in minutes.

where the birth took up to six hours, but ordinarily if more than a half hour passes in a farrowing that has been popping right along, and the sow quite obviously is trying to give birth to yet more piglets without success, something may be amiss. A piglet could be caught sideways in the birth canal, or an especially large one could have become lodged. You may scrub your arm and hand thoroughly, leave them wet and lathered, and (with trimmed fingernails) reach in carefully to see if you can help something along. If it seems there is a large piglet but one that is in line, pull it gently but firmly in time with the sow's pushing. If it is a pig coming on sideways, push it back and try to turn it to come straight on. If you cannot discover or correct the problem, call in experienced help right away.

The piglets, wet and covered by a thin membrane at birth, quickly dry and come clean without any attention whatever, especially if there is a heat lamp under which they can crawl where the temperature hovers at 85° to 95°F. (30°-35°C.).

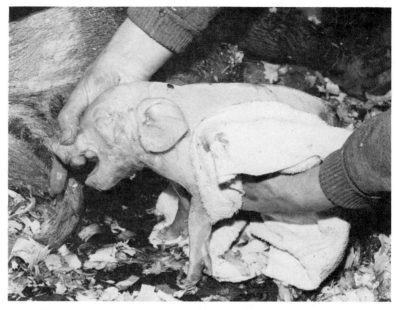

Most piglets need no attention whatever, though some farmers prefer to check each as it is born, rubbing it down with a cloth and making sure mouth and nostrils are clear of mucous.

The piglet is born looking for a nipple and that life-saving taste of colostrum.

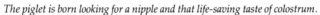

CARE OF PIGLETS

A piglet may appear lifeless at birth, but give it a chance. See that its nose and mouth are clear of mucous. A brisk massage with old cloths may bring it around. Puff some air gently into its lungs or, holding it on its back, pump the back legs rythmically back and forth until it breathes. A quick plunge in a bucket of cold water may snap it to life, or (says Youatt) dribble a teaspoon of gin down its gullet. I'd call this a last resort—dangerous and a possible waste of good booze.

A sow in the process of giving birth may be so agitated that she will attack and kill the first little ones. Or she may injure or kill them accidentally, they being ungainly and she thrashing about in the process of having more. All these dangers can be lessened by using a farrowing crate. But even so, the farmer who is concerned and who is on hand at farrowing time is likely to take up the babies as they are born, dry them off with a clean burlap sack or other cloth, make sure their noses and mouths are cleared of mucous or membranes, and perhaps put them in a box under a lamp or with a hot water bottle, keeping them there at least until all are born, dry, warm and active. By this time, too, the sow will have calmed down, or she may simply be so worn out that she is less apt to hurt her young.

If a sow does not settle down and accept motherhood, it may be necessary to tie her down or tie a burlap bag over her head for at least a first feeding session. Usually, with a bit of time and as udder swelling and pain lessen, the crankiest sow will change her tune. It is something you can really listen for—the contented "hfff, hfff, hfff," of a sow calling her young together.

Angus Rouse, on reading my suggestion that a sow be tied down, commented, "Good luck to you!" and added that from his own experiences with snappy sows, "90 percent are O.K. once they have cleaned (and so labor pains have stopped), and so leave pigs in a box till then. If the sow is still snappy, you can give a tranquilizer (*Atravet* is one brand), or if you don't have any, tie a jute bag over the sow's head, while you go for a sleep."

There is a theory that protein deficiency may drive a sow to kill and eat her young. In answer to this possibility some managers may

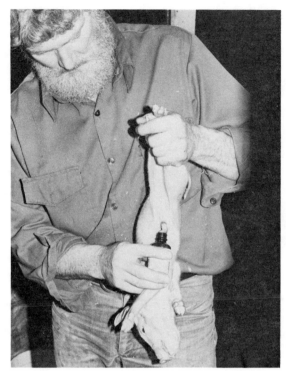

Dipping the umbilical cord in iodine wards off infection.

feed a little extra high protein supplement such as fish meal beginning ten days prior to farrowing.

Shortly after the birth of the young many farmers dip umbilical cords in an iodine or other antiseptic solution recommended to protect against infection. The cords will be four or five inches long; normally there is no need to tie them off or to cut them shorter. A small drugstore bottle of iodine makes a good applicator. Take each piglet up in turn and dip its cord completely, bringing the bottle mouth snugly against the baby's belly.

NURSING

Sows may develop individualistic styles for nursing their young, choosing to lie on one side in preference to the other, or even preferring to stand, a postion that can be awkward for short piglets.

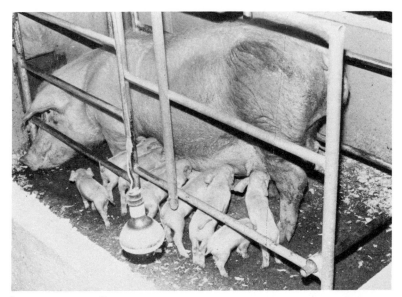

Some sows nurse standing up.

Little piglets scrap for position along mother's line of spigots. After a day or two each will have established ownership of one. In a small litter, piglets may share the nipples that are extra to those over which they have claimed exclusive rights. Nipples farther up the sow's belly give more milk and usually will be won by the larger or more aggressive piglets. Weaklings get left with "hind teat, " and so may be destined to continue the underlings in the litter.

A sow may be able to raise more piglets than she has teats—some of her babies being forced to share—but generally it will be better to take the extras away and either to raise them on a cow's milk formula or to give them to another sow. This could be a sow recently farrowed of a small litter, or a sow whose own young are ready for early weaning and who looks capable of handling a second shift. Make sure the piglets get some colostrum (first milk) before shifting them to a new mother.

Often a sow will drive off or kill youngsters not her own, but you may fool her by dabbing or spraying a smelly cologne or deoderant on her nose and on all of the piglets—her own and the fosterlings. I've also heard of using some of the mother's dung to disguise the smell of the strangers. Make the switches just at feeding (nursing) times or just as the lights are to be turned out for the night. An orphaned piglet can be raised on a bottle, but you should realize beforehand that it will take a lot of time and may end up in failure.

It is critical that baby pigs get at least one feed of colostrum within hours of birth, from its own mother or any other sow that has farrowed within the last couple of days. This gives the youngster disease-fighting antibodies it is not born with and which it will not be able to produce on its own for a week or more.

After having been fed a shot of colostrum the orphan may be reared on a cow's milk formula. Feed it at body temperature in a bottle with a small nipple, from a medicine dropper, a cotton swab soaked in the formula, or by spoon. Especially for the first week, a tablespoon or so of milk at frequent intervals will be needed—six, evenly spaced feedings a day, or more if you can manage. Trying to feed large quantities of milk at one time will lead to scours. A piglet can learn to drink from a saucer in a matter of days, and the formula can be fed at room temperature after a week or 10 days.

MILK FORMULA FOR ORPHAN PIGS

1 pint whole milk
1 tablespoon skim milk powder

(This formula comes from Ensminger. Older books often recommend adding two teaspoons of glucose or corn syrup, and two grains of sodium citrate may be recommended as well. However, as the newborn piglet's system is predominantly a milk-sugar (lactose) digester, Ensminger's opinion that other sugars not be included makes sense.)

WEANING

Piglets may nurse for 10 or 12 weeks, but they can be weaned as early as 3 if they have been well started on alternative foods. Normally, nursing sows don't come into heat, and so the aim of commercial producers is early weaning in the interest of getting mothers bred back as quickly as possible. Three weeks (12 to 15 pounds) or five weeks (15 to 20 pounds) to weaning are the programs most often followed, with the piglets "started" on a commercial medicated feed or they may be started on a non-medicated, home-mixed ration having 20 to 22 percent protein.

Weaning is done rapidly. Usually the sow is taken away and beyond earshot of the babies, who may be left in the farrowing area for a day or so to minimize their shock. Water may be withheld from the sow and feed reduced for a day to help check milk production.

A Place to Farrow

PENS OR CRATES

If you have a number of sows, it will pay to have certain pens or crates reserved for farrowing. Expectant mothers should be placed in them a week to three days before they are due, to let them grow used to the new surroundings. There should be one farrowing pen or crate for every two to three sows in the herd. One of the arguments for early (three-week) weaning is the fewer farrowing units required.

Homemade farrowing crate

These pens or crates must be cleaned between uses. Scrub walls and floor with a lye or other cleaning solution (if lye, mix a pound to 15 gallons of water) and later flush with a disinfectant (from a livestock supply house or other source), and allow to dry.

The sow or gilt headed for her sparkling delivery room may be washed and hosed down also, to remove parasites and their eggs. This does not totally prevent contamination of the babies from the likes of round worms, but it helps.

BEDDING

Chopped hay or straw, coarse sawdust or shavings make excellent bedding materials, but not too much and definitely not long hay or straw, which may cause piglets to be tripped up and caught beneath their mothers. You may give long hay or straw to a good mother in a pen a day or so before she is due. She'll chew it up fine for a nest. Remove any she has left long.

A farrowing pen should be six-by-eight feet in size and have piglet guard rails at least along the sides. These may be planks or poles 8 to 10 inches off the floor and reaching about a foot out from the walls. This kind of arrangement pretty well assures piglets won't get squashed against the walls as their mother lies down or rolls over.

Protecting the piglet.

With farrowing pens especially, piglets need a creep area. Heat in the creep will encourage the little ones to snuggle together and out of harm's way beneath the sow. If you haven't power in the barn and don't want to have the piglets in the house those first cold nights, at least give them hot water bottles wrapped in feed sacks and placed in a low, snug creep or box within the creep area. Make sure heat lamps cannot be reached by curious snouts; they are very hot and can cause fires if they fall against dry bedding.

Some of the advantages of farrowing crates are saved space and bedding—the sow can only excrete in one spot—built-in protection for the young, and their confinement of sows, as this facilitates taking temperatures, giving injections, and so on. Disadvantages are lack of exercise for the sow and the more limited chances for the piglets to learn about eating or cleanliness from her.

It seems the confining nature of a crate is not as cruel as it would be, say, with a dog or cat, since sows do not groom their young or otherwise give them nearly as much attention as is common throughout the world of furry land animals.

One attention a good sow does at least attempt to pay her young is to help them find her udder. She roots her piglets back at feeding time, or while on her side, sweeps back with her fore-trotter, clearing the youngsters away from her face and neck and back toward her udder.

There may be many other things sows do for their young that we don't notice. Certainly there is important communication going on through grunts and squeals. Remember this when you have to work on piglets, so that their squeals don't upset the mother too much. It may be smart to take them into another room before nipping, notching or cutting operations. Here again, though, the danger of a sow hurting herself or others of her brood because you have made one of them squeal is greater in a pen than if she is in a narrow farrowing crate.

(See Chapter 7 for precautions in laying a floor beneath a pig that is being kept in a narrow stall or crate.)

Culling Sows and Boars

There is no hard line on what age to cull a sow. She may produce two good litters a year for six or more years running, or she may not be worth taking past her first farrowing. Morrison says "sows which produce especially good litters and which are good milkers and careful mothers should be retained as long as possible."

Fishwick suspects a relationship between numbers of pigs crushed and the ability of a sow to produce milk. He feels older sows may crush more young not so much because of aged clumsiness as because the piglets are being starved and weakened due to lower milk production and so are less able to scramble from danger.

On an average, gilts have fewer piglets than mature sows, but some of this slack may be taken up by making sure not to breed the gilts at too early an age. A gilt farrowing only five or six piglets may be kept for a second try if she has the pedigree and traits you like.

But if, as a mature sow, she gives you no more, she probably should be culled. Commercial operators in Canada figure they can only break even with a sow that *successfully weans* a minimum of 12 piglets a year.

RECORDS

Keeping records of heat periods, breeding and due dates, litter numbers, dates of weaning, and numbers weaned will make all the difference in your operation, no matter how small it is. If nothing more, hang a large calendar in the barn, with a pencil chained nearby, for noting these and other important figures and events.

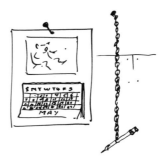

CHAPTER 14

Management Techniques

Most of the techniques covered in this chapter have to do with nursing piglets (up to three or five weeks of age, depending on the management system being followed) and will not be part of raising a weaned feeder pig to market weight. Exceptions may be *weighing, ringing, slapping, mixing* and what I'm calling *housetraining*— techniques to keep confined pigs clean.

First Steps

KEEPING TRACK

In the course of carrying out many routine procedures with a litter of pigs, a lot of time will be saved if, as each one is processed, it is well marked in any way that tells you, "that one's done." A bold felt pen, soft crayon, grease stick or soft chalk will serve to leave an obvious slash of color across a pig's back.

CLIPPING NEEDLE TEETH

As mentioned earlier, piglets are born with eight tiny teeth that are well named "needle teeth." Sometimes these teeth scratch a sow's teat, and sometimes arguments over a teat result in piglets dealing each other scratches and nips that may become infected. The answer

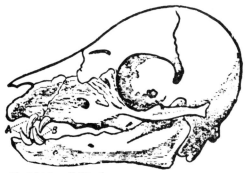

The piglet is born with eight "needle" teeth.

to this on most commercial farrowing operations is to simply clip the tops of all needle teeth shortly after birth. That's *all* of them, not just those that seem abnormally long or that are dark colored.

People raising piglets from a few sows may not want to clip needle teeth, and perhaps this is better, because teeth are ocasionally broken when clipped, and abcesses or gum infections can follow.

The tips of "needle" teeth may be removed to lessen injuries to the sow's teats.

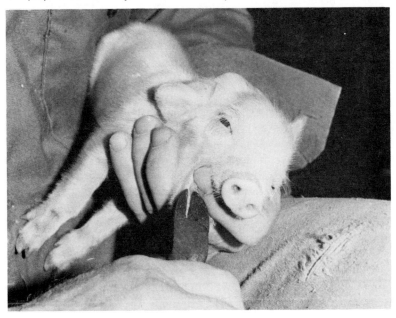

A middle-of-the-road approach is to check the mouths of new piglets and if any seem to have unusually long teeth, deal with them alone. If a sow's teat becomes inflamed, check the teeth on the piglet that has assigned itself to that nipple. Don't follow the advice in one old book, which was "If a sow's teat is so sore she will not let the pigs suck, cut it off and save the pigs."

Use wire nippers or toenail clippers to clip teeth. Hold the piglet's mouth open with your finger far back between the jaws. Only take the points off the teeth.

IRON SHOTS

Piglets are born with a low reserve of iron, which is the mineral needed by the oxygen-carrying hemoglobin of the blood. They don't get iron in their mother's milk, and so this is the first "nutrient" they must get for themselves, and soon. Anemia, which results from a shortage of iron, can set in the first week of life. Piglets born outdoors may be able to get their iron by rooting for it in the dirt and sod. But if they are being raised on a wood or concrete floor they are in a bind.

There are several solutions: one to throw a shovel of dirt or sod to the piglets two or three times a week so that they may root for iron in that, another is iron tonics given orally. A solution of ferrous sulfate (1 lb. in three pints of water) may be painted on the sow's udder daily until the piglets are eating plenty of solid foods, but this is time-consuming and messy.

The most popular method, probably because it is quick and done in one operation, is to give an iron injection in a ham some time within the first three days after birth.

Some people give the injection in the "fleshy" part of the neck—it is given into muscle rather than directly into the bloodstream where it would likely prove fatal—because they fear staining the meat of the ham with the iodine-colored iron. But staining is rare, and I think for novices greater danger would be the chance of shooting iron into the neck vein of a squirming piglet.

Shooting into a ham is a simple, one-person job. Hold the piglet by the leg that is to be injected. The thumb of that hand can push the skin to one side at the point where the needle will be inserted. In this way when the skin is released following the injection the needle

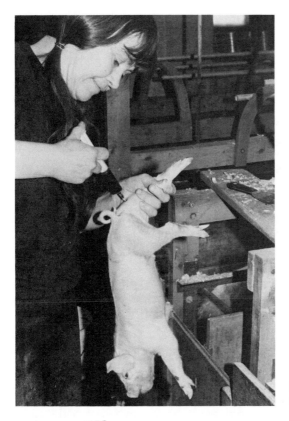

Iron shots prevent anemia.

hole in the skin will not line up with that in the muscle, and little if any of the iron solution can escape.

A 1 or 2cc. injection of iron solution is a typical dose, the amount depending on the preparation used. (See instructions on giving shots in Chapter 15.) Over the days and weeks that follow until the piglets are on their own feeds, watch for signs of anemia (Chapter 5) and repeat dosage for any that look in need.

EAR NOTCHING

Whenever numbers of pigs are raised, it becomes increasingly important to keep track of who came from what and when. Notching ears is the oldest, easiest and cheapest method, and the obvious choice for small farms.

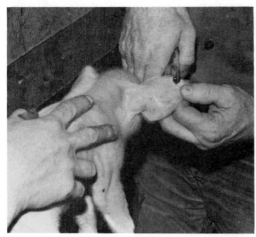

Ear notches are an easy-to-read identification mark.

There are several different notching systems, and if you have purebred registered stock, find out if there is one preferred by your breed's association. Otherwise you could as well make up your own system or adopt whatever system is popular in your area.

One of the most popular for identifying up to 161 litters gives the litter designation to the pig's right ear and designation of the individual within the litter to the left.

In this system each ear is divided into outer and inner halves and top and bottom edges. There can be one or two notches in each quadrant. In addition, the right ear's tip is notched after the 80th litter to begin again with number 81.

Notching.

Notching is done early in the piglet's life. It may be done at the same time as one of the other procedures—teeth clipping, docking or such—but it is best not to combine any more than two procedures. More would put too much stress on the piglets.

A leather punch, razor blade or special ear notching tool may be used to cut the ears. The notch should be about a quarter of an inch deep. It will end up far larger and be easily read on the full-grown animal. With this in mind, when notching the ear of an older animal make the cuts correspondingly larger since they will not be expanding to the same degree.

TATTOOING

Tattooing ears is for big producers of registered stock, and is done before weaning. It is a simple procedure. The tattooing machine, a costly tool, is like a pair of modified pliers in the jaws of which letters or numerals made in a relief of needle points are inserted. An area in the center of the ear between two major blood vessels is punched with the selected numbers and/or letters. Ink is then rubbed into the pin-prick wounds.

Reading tattoos can be difficult, especially on dark-skinned pigs. It may help to shine a flashlight through the marked ear.

DOCKING

Crowded feeder pigs being fattened for market are like work-bound commuters on a subway. They have to chew something. It's nervous tension I guess, which might not be a problem if pigs could chew bubblegum. But they can't, and so they chew each other's tails. If a piglet would leap away when it felt a tooth on the tip of its tail, open wounds and crippling infections might not follow. Apparently, though, there is little feeling there, and a piglet may stand nonchalantly at the feed trough, munching away while its flank is under attack.

Until quite recently the standard practice on farms where tail biting was a problem was to dock the entire tail. Now more farmers are docking only the least sensitive tip. It is done while the piglets are tiny. When carried out with a minimum of fuss and excitement

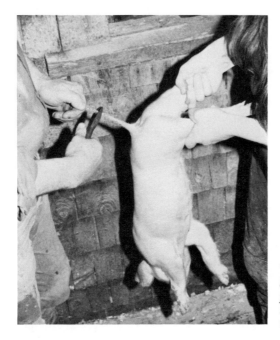

Tail docking reduces tail biting among closely confined pigs. Only the last, least sensitive third of the tail need be removed.

there should be little bleeding. Wire nippers used to clip needle teeth do a good job. Snip off the last,least sensitive third of the tail, cutting down in a groove between vertebrae. The wound may be dobbed or sprayed with a disinfectant preparation recommended for cuts and open abrasions.

No docking may be necessary on a farm where there is plenty of room for the pigs and where there is time to assess and correct conditions that have led to biting, and to notice and treat wounds.

Castration

Because of the trauma, it is usually recommended that male piglets be castrated while they are still nursing, and preferably four or five days before they are to be weaned. Initially this may pose a problem for new farmers hoping to wean their piglets at three weeks because at age two weeks the little male simply hasn't much to work with. However, with practice the problem will dissolve. In fact, the earlier castration is done the better it is for the piglet. Smaller wounds accompany the operation, healing is rapid, and trauma is minimized.

To castrate a male piglet you usually need two people. The holder places the piglet on its back in his or her lap with the rump facing out. Hands hold the piglet's hind legs forward. This presses the testicles tightly against the scrotum while at the same time confining the piglet. Swab the scrotum with a disinfectant solution.

The person doing the cutting presses the testicle against the skin on one side of the scrotum between thumb and forefinger and cuts down with a sterile razor blade or scalpel. One cut. Don't saw. Pressure behind the blade should assue that the cut is through to the testicle. It may go into the testicle but there is no harm in that. The cut should be a half to an inch long and low on the scrotum to assure proper drainage of the wound. Remember that "low" will appear high on the upturned piglet.

As soon as the cut is made, press the testicle out and pull it gently away from the piglet's body. It will be attached by what at first looks like one piece of stringy tissue. There are basically two running together. One is the sperm duct and is whitish. The other is blood vessels and is reddish. The whitish cord may be cut. The blood vessels should be snapped off as long as possible and close to the body. To do this merely keep pulling the vessels out sort of hand-over-hand on a mini-scale until resistance from within causes the tissues to pull apart. Sounds gruesome, but it is by far the best way because as the artery snaps, the remaining portion shrinks back inside the body, closes off, and bleeding is at a minimum—much less than occurs when the vessels are cut.

The process is repeated on the other side of the scrotum. The wounds are treated in the same manner as that following docking. No stitches are needed.

If you have to castrate alone, you can buy a pig holder. Or you could build one. All that is needed is a "V" trough with straps for holding the piglet down and hind legs forward. There is also a way for one person alone to both hold and castrate a piglet, but it is not easy with one that is under a month old. It takes a longer piglet whose head and snout can be clamped beneath one's leg cocked up on a chair behind, and whose testicles are easier to find than are those on a piglet only two or three weeks old.

Following any surgery there is a danger of infection from flies

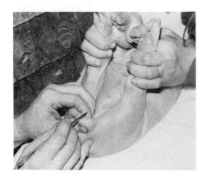

Castration: A. After swabbing the surface with a disenfectant, the testicle to be removed is pressed against the scrotum. Cut once. Do not saw, as this can create a more jagged wound that is slower to heal.

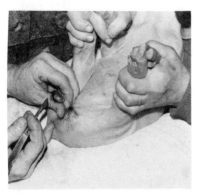

B. The testicle is popped through the incision and is drawn out from the body.

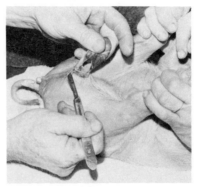

C. Two cords, one white and the other red, link each testicle with the abdominal cavity. Cut the white cord (seen on the left.)

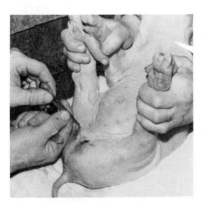

D. Draw the red cord containing the blood vessels slowly out until it snaps. In this way bleeding is minimal.

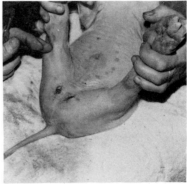

E. With practice, castration takes no more than a couple of minutes. The wounds are small and clean and low on the body to assure drainage and rapid healing.

looking for places to lay their eggs, or from bacteria. The best way to avoid flies is to operate in cold weather. Some old books suggest putting ashes in a wound. I suppose this would help staunch a flow of blood, and if they were fresh they would be cleaner than dirt. Staunching bleeding is the object of packing castration wounds with baking soda, which is another practice still common today in some areas.

The surprising thing is how little bleeding there is and how seldom complications such as infections set in even when nothing is done to treat the wounds. In fact many recommend not packing wounds or applying strong disinfectants of any kind, as these practices may cause more complications than they are intended to cure. More important is placing the castrated pigs back in a clean, freshly-bedded pen. Some swelling is to be expected following castration, but there is no need for concern as long as the piglets are active and eating.

With a little practice, castrating male piglets is very fast and creates little trauma in either the piglet or the operator. On being released, piglets scramble back to the litter for consolation and instantly nurse if the sow is willing. Amazingly they will be acting as though nothing has happened within minutes of the operation.

CASTRATING OLDER PIGS

There are times when an older male needs to be castrated. Perhaps it is a young pig kept on to see how it might shape up as a breeder. Or it could be a five- or six-year-old boar no longer suited for breeding. In any case, if the animal is more than a couple of months old, the cords should all be tied off with surgical or other strong thread—dental floss will do in a pinch—and then cut. Otherwise bleeding could be excessive. On the oldest animals the shock could be tremendous and it is reasonable to call in a veterinarian so that the boar can be at least locally anesthetized.

Once castrated, they are called *stags*. The operation should come five to six weeks before slaughter. Some prefer to castrate a boar and then put it on pasture for a summer before slaughter. Others will thin a boar down before operating, then, three or four days after castration when the stag begins to find his appetite again, they start pushing the grain to him and continue a liberal feeding program for two to three months.

Long says, to restrain a boar for castration, noose its nose, fasten the noosing rope to a solid post, and rope up one hind leg.

I asked one farmer who castrates several boars every year how they do it on his farm. He replied, "fast." He followed by saying they do noose the boar's snout, secure the rope, and then one person forces the boar against a wall, knees braced against the animal's side, while the "surgeon" cuts each side of the scrotum in turn, pulling the testicles away until the cords snap. They do not tie and cut the cords on these four- to six-month-old boars that did not measure up for breeding stock. Neither do they anesthetize them. All efforts are, as the man said, toward making a speedy job of it.

FEMALES

At one time female piglets were castrated, but this was when pigs were grown two or three years before slaughter. Zeuner wrote, "the female piglet is castrated by means of two cuts about two centimeters long through which the ovaries are extracted with a small hooked tool and then pinched off with the fingers. Without any further ado the piglet is allowed to run away."

Another practice for inducing growth in female pigs comes from Long, who wrote, "Some persons prefer to have young gilts served instead of cut.... This gives an impetus to the feeding propensity of the animals, and to the laying on of flesh, and if they are killed in five or six weeks, no ill effect will result, for the immature state of the young will have caused little or no drain upon the system."

Managing Older Pigs

WEIGHING

Piglets can easily be weighed to see how they are progressing by use of a simple, hanging spring scale. Place the piglet in a sack or box that can be hung from the scale, or use a large funnel in which the

piglet can be placed, snout down. Or you can hang the pig on the scale by means of a strap slip-knotted over a hind trotter. For pigs of more than 50 pounds you probably will want a holding crate and a platform scale. Proper large pig scales are available for a price.

The weights of larger pigs or hogs can be estimated from measurements of heart girth and length, or from heart girth alone (see the accompanying table). Also special tape measures for estimating hog weights are available from livestock stores or supply houses.

One formula for estimating weights worked out at the University of Minnesota goes like this:

Heart Girth × Heart Girth × Length ÷ 400 = Est. Wt. (lbs.)

The *length* is the distance from the base of the tail to a point mid-way between the ears—with the pig's nose off the ground. The *heart girth* is taken just behind the front legs. The people who worked out the formula claimed fair accuracy for hogs up to 400 pounds.

Measuring the pig to find its weight.

FAT PROBING

Fattening is a highly heritable characteristic in hogs and is credited half-and-half to influences of the boar and the sow (sire and dam).

Taking back fat measurements is either costly or painful to the hog. It isn't something you do out of idle curiosity and it is usually reserved for selecting breeding stock.

The degree of fattening overall can be judged from measurements of the outside layer of fat covering a pig's back and loin. Today the measurements can be taken with an electronic, sonar-like instrument that shows at what depth the denser muscle layers lie beneath the skin and fat. But these machines are expensive, and unless a government or association provides a back fat measuring service most farmers fall back on the older system that requires poking a measuring probe down through the fat.

As mentioned before, the preferred market hog today has about an inch of backfat when it reaches a live weight of approximately 200 pounds. The potential breeder should be better than average.

When a fat probe is used to measure the backfat, from two to six measurements may be taken; the more to assure no mistake is made due to a faulty job of probing.

Sites for probing.

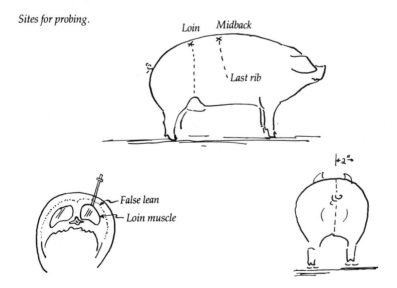

The animal being probed must be held with a noose over its snout. The sites for probing are over the mid-back and loin, one to two inches to right and/or left of the midline. First a small incision is made in the skin with a scalpel taped to prevent too deep penetration.

The incisions are made *across* the long axis of the pig. Then the probe, held vertical to the surface of the skin at the site of a wound, is pushed down through the fat until it meets solid resistance from muscle layers beneath.

There is a layer of connective tissue about halfway to the muscles called the *false lean*. The probe can be popped through this layer with fair ease. It is the chance of accepting a "reading" at the false lean that encourages the beginner to repeat each probe on either side of the midline. With a little experience, however, it will be enough to take two probes.

Following probing, the figures are averaged to get an overall measurement for backfat that will be used to compare the form and performance of one to pig to another.

Although infections following probing with a sterile scalpel and probe are rare, it is good pratice to swab the probe sites with an antiseptic preparation before and after the procedure.

Besides being painless, electronic "probing" is faster than the mechanical method. Too, the electronic probe can be used to measure the depth of the muscles lying beneath the fat at the loin, giving readings from which the loin "eye" area can be determined.

FOOT TRIMMING

Occasionally a sow's or boar's hooves grow too long and must be trimmed back. Ensminger says the frequent culprits are the outside toes on the back feet, which "grow faster than the inside toes." He suggests coaxing the animal onto its side by scratching its belly to get at the feet with hoof shears, pruning shears, a knife or rasp. The job takes a lot of patience, especially if the pig gets frightened. It may help to have a helper to continue scratching the pig's belly while the trimming proceeds.

Only take off a little extra hoof at a time, sticking to the shell-like excesses growing beyond the oblong cores of each toe.

RINGING

Pigs are ringed to stop their rooting, fighting, fence-breaking, gate lifting, and to help break up the parasite cycles that include intermediate hosts such as earth worms that live in the soil. Farmers who have provided improved pastures for their hogs hate to see them plowed under. The answer is to put rings in pigs' noses.

Ringing is done first when pigs are small. It is a traumatic experience and so should not be done in conjuction with any of the other painful piglet procedures. In the same vein, ring sows when they are "open" rather than when they are pregnant. On permanent stock rings may have to be replaced every several months as the old ones wear out.

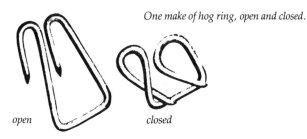

One make of hog ring, open and closed.

open   *closed*

A bent nail can be used as a ring, but store-bought rings don't cost that much. If you can't find them locally try one of the livestock supply house catalogs listed in the Appendix.

A piglet can be held in the lap for ringing. Larger pigs will have to be held with a snout noose or in a stock. There are two ways to ring a pig. In one, a large ring goes through the septum (nostril divider) about half the diameter of the ring in from the outer edge of the nostrils. The other way is to place three or more small rings in the upper rim of the snout.

Another method more common in the past to stop rooting was to cut through the skin and ligaments connecting the top of the "floating" disc of the rooter to the snout proper. This operation would be performed on weanling piglets. The skin would heal but not the ligaments. The rooter was left powerless.

And still "another plan," writes Long, "is to thrust the point of a

A snare or holder is needed to restrain pigs for ringing.

penknife through the rim of the snout and to draw it half an inch to the right and left so that as the pig roots the piece of loose flesh will give sufficient pain to cause him to desist."

SLAPPING

Twenty years ago a hog slapper was a rig for driving pigs—a strap of rubber or leather fastened to an old paintbrush handle and used to whack balky animals roundly and most of all loudly. Today a *slapper* is a tattooing instrument used on pigs headed for market. Some *Lot #* or other significant message is placed in the inky slapper, which is on a hammer-like handle. As the pigs are rounded up for market, each is slapped on a shoulder, leaving a mark that can be read for several days, even after the pig has been slaughtered and the carcass scalded and scraped.

MIXING PIGS AND STOPPING FIGHTS

Pig fights are not always amusing tests of will. They can be deadly when several pigs decide to gang up on one of their number. The worst break out when one or two pigs are added to a large group in which the pecking order has become well established.

Newborn piglets can get into nasty little scraps that can be a problem when attempts are made to even the litter load among two or more sows that farrowed within days of each other. For more about *fostering* see Chapter 13.

Some thoughts to keep in mind when pigs must be mixed are:

1. Don't put small pigs in with larger ones. Try to keep all groups segregated by size.

2. It is best to mix large pigs outdoors or in a pen that is new territory for all concerned.

3. Hold off feeding pigs to be mixed and feed them as soon as they are thrown together. Don't feed pigs and *then* mix them up because then if fighting follows the situation will be aggravated by animals vomiting. A gut distended with food will be more susceptible to damage from the pummeling pigs deal each other with their snouts. In fact it is unseen, internal injuries that often lead to death. There won't necessarily be massive, visible wounds.

4. Some recommend mixing pigs at night and leaving them in the dark.

5. Don't mix one or two pigs in with a larger cohesive group.

6. Mix several pigs of roughly the same size. The best would be to mix a few pigs from each of several litters.

7. Spray the pigs with paraffin, cologne, linseed oil or anything else that will make it hard for them to tell difference by smell.

Pigs crowded together have shorter tempers. Tails get chewed so tails get docked. Then ears get chewed, and maybe it's only a matter of time before breeders work for pure breeds of earless porkers for the commercial feeder market.

Someone discovered that if pigs in a tight place have something to play with they fight less. So now a length of heavy chain hanging from a ceiling into a feeder pen is a common sight on many hog farms. It doesn't have to be chain. Almost anything hanging down into the pen just about to the floor will do; something to spar against like a punching bag.

Even under the best management fights will break out. They don't always indicate a need to move pigs again. Maybe the pigs are only getting a new pecking order down pat. Yelling at the pigs may stop a row. So may throwing a couple of tin cans into the pen, some

A chain or tire hanging down in a pen of pigs provides a trouble-free outlet for aggression.

hay or (one of the best distractors) a paper feed bag. The pigs tear into the bag. The bag flaps and snaps and hisses back, and the belligerent pigs apparently are as well rewarded as they are by a good fight.

DEFENDING YOURSELF

Occasionally you have a sow that gets nippy, barks at you when you go in the pen, and you'd get rid of her but she's an exceptional animal for some other reasons. Or you have a boar and you simply would feel more comfortable dealing with him if you knew how to defend yourself in the face of any displays of aggression or whatever an advance portends.

In all cases the best protection is caution. It may sound foolish but in fact it's when you are in a hurry and put caution aside that trouble is most likely to pop up.

Boars must be respected, even if they are always calm and gentle. Some handlers will never enter a boar's pen without taking one of those light, hand-held hurdles. But another handler I spoke with feels he is better off having his hands free to speed his vaulting over the fence if the need arises. He used to carry a hurdle until the day a boar twitched his head and, SHAZAAM!—the hurdle was gone. "I had a good grip on the handle," recalls the handler. "But that's all there was left."

A hog slapper of the old, noise-maker variety, could straighten out a snippy sow. Or Long says to use a "short, fine ash stick...on the snout of an unruly animal."

TUSKING

At 18 months to two years it is a good idea to cut back a boar's tusks. This is done with a file, hack saw or hoof nippers while the boar is secured by a noose over its snout.

HOUSE TRAINING

The pig in a pen will usually establish a manuring area very quickly, and if there is a low or moist spot—especially if these two are together—that's where it is likely to be.

If pigs that are usually good about manuring a particular spot in the pen start messing up, see that they are not overheated. Try cleaning all manure away daily *except for one portion* left in the proper manuring area. Make sure the pen has plenty of ventilation.

CHAPTER 15

Health

Cleanliness is next to healthiness. This is not only because germs like dirt, but because we people don't like animals that are caked with mud and manure. If we let pigs get filthy, we spend less time caring for them. They go from filthy to scrimey, and soon the porkers are getting but a few begrudged minutes a day, coming at them in a slop bucket. The stage is set for disaster.

Managing for Health

A well-fed and well-housed pig can cope with most diseases, and it is not likely to suffer injuries. The best health program, then, comes with proper management. From the opposite direction, having discovered sickness or injury, don't treat symptoms without looking further to see if there is not a cause lurking in faulty management.

LOOK FOR PROBLEMS

Find a cut? Look for nails or other sharp protrusions. Pig got a cold? Look for drafts. Look for a dietary deficiency. Look for both. Got a lame pig? Again look at diet. Look at your floors. And so on.

Looking is so important—*every day*. Once a germ strikes a growing

A good environment for diseases and parasites.

pig it works its evil wonders with incredible speed. Maybe this is tied to the amazing rate of growth. Certainly related to growth rate is the speed at which a nutritional deficiency can develop to a crippling degree.

A good swineherd keeps pigs in clean, well-lit quarters and watches to see they are all eating and drinking, watches to see they move normally, that there is no limping, staggering, circling or shaking.

Look for all those signs of a healthy pig that are mentioned in Chapter 5: stools that are well formed but not dry, a general alertness and interest in the world around, a body well fleshed but not fat.

A pig may lose its appetite for a day, but any longer suggests something is seriously out of kilter. Take the pig's temperature with a rectal thermometer. Insert the tip as deeply as possible in the rectum—slid in gently with the aid of Vaseline or a bit of mild, soapy water. It is a simple task to take the temperature of a pig that is busy eating. But if it is not eating, and won't stand still, you may have to crate it up, tie it down or snare it.

A temperature of 102.5°F. (39.2°C.) is average among healthy pigs, with a range of 101.6 to 103.6°F. (38.7 - 39.8°C.) being accepted as normal *(Duke's Physiology)*. Therefore a temperature of 101 or 104° F. will be a clear indication of trouble, especially when coupled with any abnormal look or behavior.

INTERNAL PARASITES

A cleanliness not so easily seen but terribly important to maintain is *within* the pig. Internal parasites, especially worms in the stomach and intestines, can keep you broke, destroy the health of the pig and lead to sickness and death.

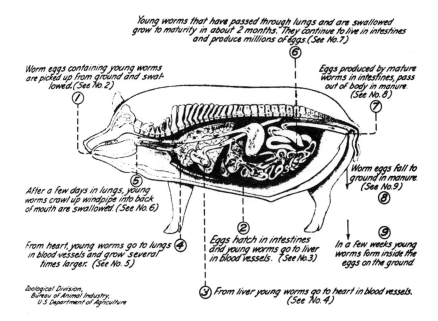

The Roundworm's Journey Through the Hog: The course traveled by the roundworm in the pig. The worm eggs are swallowed by the pig and hatch in the intestines; the young worms go by way of the blood vessels to the liver and then to the lungs; here they leave the blood vessels and enter the air passages, go up the windpipe to the mouth and are swallowed; return to the intestines where they develop into adult worms. The female worms produce eggs which pass out in the manure and start a new journey through the pig that swallows them.

If you buy piglets, they should be wormed immediately and placed in a quarantine pen for a day before moving them into permanent quarters. With some medications a second worming ten days to two weeks after the first is recommended to assure destruction of eggs or larvae that were not killed in the initial treatment.

Many people raising pigs establish a routine whereby all breeding stock gets wormed at least twice a year. Boars may be wormed any time. Sows should be wormed prior to breeding or toward the latter stages of pregnancy.

If you are pasturing pigs, the ideal is to worm just before turning them onto new ground. Pastures should be rotated so that fields the pigs have been on and infesting are given vacations lasting as many months as possible, so that larvae from worm eggs deposited in the manure may starve. There is an ideal of from one to three years before pigs are returned to an old pasture but this can be hard to manage if you are stuck with limited space and fencing. A period of cultivation helps to clean parasites from land that has been pastured.

GUARD AGAINST OUTSIDE DISEASES

Those who raise a pig or two each year risk little in the chance that a visitor—pig or person—will bring disease into the home farm. But when you have breeding stock and/or large concentrations of pigs the doors should be closed to casual mixing between outsiders and your livelihood.

Most large hog farms try hard to keep their animals isolated from infection. Indoor piggeries even wire their windows to keep out wandering dickey birds. Barn doors are kept locked, and necessary visitors are given plastic booties or are asked to walk through disinfectant shoe baths before entering the pigs' homes.

New pigs are brought into a breeding barn as seldom as possible and only after they have been kept in quarantine from the rest of the herd for at least 30 days. An extension bulletin from one of the southern states suggests building the isolation barn *downhill* from other barns and pastures so that run-off won't pass under the main herd.

Giving Medication

It helps—and may save money—to know some of the routine techniques for delivering medicines to sick animals. I'll stick to a few of the simplest and most routine here. If you want more, perhaps your veterinarian will take you under his or her wing for a visit to a hog farm. An hour spent with an interested vet would give more information than I could hope to deliver in chapters.

ORAL MEDICATIONS: PILLS, POWDERS
AND DRENCHES

It isn't hard to pry open the mouth of a baby pig to give it a pill, placing the tablet as far back in the mouth as possible so that it has to be swallowed. With larger animals you may need to use a *balling gun*, which you can make at home out of a length of garden hose or plastic tubing (about 18 inches long) with a plunger of smaller diameter hose or a wooden dowel inside. The pill or *bolus* goes in the end of the larger, outside tube, the tube is worked far back in the pig's mouth, and the plunger pushes it home. A larger pig will have to be snared to give it a pill.

Some dry drugs may be eaten if they are in a pig's water or are mixed in the food. A vet here suggests turning bad-tasting powdered drugs into irresistible treats by mixing them with dry fruit jello crystals.

Drenches are liquid drugs given by mouth. Little piglets may be held by hand for drenching. Larger ones may have to be snared. Give a drench slowly and don't hold the snout higher than the pig's shoulder. Ignore these pointers and risk drowning your animal.

INJECTIONS

There are four types of injections named for where the tip of the needle delivers the medicine. Different locations are chosen depending on the type of medication being injected, where the medicine is to do its work, and the speed needed or desired in getting the medicine where it is needed. All sites must be cleaned and disinfected.

An *intravenous* (into bloodstream) injection delivers a drug speedily to all parts of the body. But some drugs would cause clotting, possibly leading to death, if they went directly into a large vessel. These, then, must be injected into the body cavity *(intraperitoneal)*, a muscle *(intramuscular)* or under the skin *(subcutaneous)*.

The right length and bore of needle is important. It is always best if you can insert a needle to its hub. As an extreme example, imagine what would happen if you were trying to give a subcutaneous injection with a two-inch long needle and the pig jumped. There's no telling where the injected drug would wind up.

There are two ways to avoid the hazard of delivering a drug directly into the bloodstream. One is to insert the needle alone and to wait a moment to see that blood does not drip from the hub before attaching the loaded syringe.

The other way is to insert the needle and syringe as one, and to draw back on the plunger before shooting in the drug. If, on drawing back, blood appears in the barrel of the syringe, the needle is withdrawn and another site is chosen. This second method is the one I first learned and used for years, but on reflection I see it has its drawbacks for beginners working with squirming animals. Too, in the anxiety to see the job done, I have seen that crucial pull-back-on-the-plunger step forgotten.

Pigs take intravenous injections in the ear, in one of two prominent vessels that branch from the base of the ear outward toward its tip. A firm but not crushing pinch at the point of branching causes the vessels to distend for easier insertion of the needle.

Intramuscular injections should go into areas of heavy muscling, such as the hams, the shoulders and the top of the neck. As a rule, no more than 10 or 15 ccs. (cubic centemeters) of a drug should be injected into a muscle at any one spot at one time.

Intraperitoneal injections go into the abdominal cavity in a piglet held up by its hind legs, or a larger hog lying on its back. Choose a site below and to the right of the navel (corresponding to the pig's right flank).

To give a *subcutaneous* injection, pinch up a fold of skin and insert

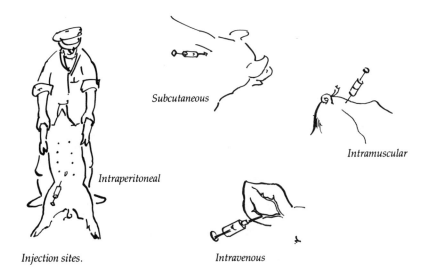

Subcutaneous

Intramuscular

Intraperitoneal

Injection sites.

Intravenous

the needle at a flat angle so that the drug goes into a small pocket between the skin and the underlying muscles or fat.

Although the best policy with disposable syringes and needles is to throw them away after one use, many of us penny-pinchers find that we can use them over again if they are properly handled. They should be washed and rinsed repeatedly by drawing and ejecting fresh and soapy water. Needles, plungers and barrels should then be separated and, along with plastic needle guards, be immersed in boiling water for at least ten minutes in a little pot kept for the purpose (boiled plastics sometimes give off an odor and flavor that may be picked up by the pot). Dull needles may be sharpened on a wet stone or emery paper. Syringe barrels and plungers should be stored apart. Otherwise they can become stuck together.

TOPICAL ANTISEPTICS

Under the best conditions pigs will get scraped, cut or bitten, and any break in the skin is a door for germs to enter. One of the handiest pieces of first-aid equipment is a bottle or spray can of antiseptic kept at the pigs' home so that germs are given a licking before they get established.

When to Call the Vet

Whenever an animal shows signs of illness, one of the first questions is whether it is serious enough to call in the veterinarian. Can we afford to wait, on the chance the pig will either recover (from a mild sprain or upset) or at least develop more readable symptoms?

Unless you have the training to deal with the problem yourself, you should call a vet right away if:

1. The pig is down and cannot get up.

2. The pig is staggering about or shaking uncontrollably.

3. The pig has not eaten for more than a day.

4. The pig has an abnormal temperature (below 101°F. or above 104°F.).

5. The pig has a wound that is bleeding profusely.

6. There are indications—either through a foul smell or inflamation—that a wound has become infected.

7. The pig has developed bumps or lumps.

8. The pig is suffering spasms of choking, coughing or sneezing.

9. Anything else either in the look or behavior of the pig that has you worried. Remember that the sicker the pig becomes the more costly the situation may be for treatment and in terms of lost growth—also, of course, the more risk of losing the pig.

Often pig cures are simple and inexpensive. For instance, we had a 10-month-old gilt in the barn that developed a lameness in all four feet in less than a week. Then suddenly she became paralyzed and could not get up without help. We had been attempting to find what was causing the lameness but now it apparently was too late. We would have to do her in. First, though, we called the vet on the off chance he would have a miracle cure in his bag of drugs. He did have a cure, but without drugs or cost. "Get her outside," he ordered. "Right away."

Severe anemia set this pig back from littermates.

We did as instructed, moving the gilt to an outdoor pen on sod ground. Within three days she was a new animal. It may have been the change of footing that made the difference. The barn floor was rough. But I suspect what brought our pig around was the better diet that came with grazing and rooting, the direct sunlight for its vitamin D, and the exercise.

Another incident, this time involving a neighbor, reinforces the first argument that good food and housing grows healthy pigs. In this case two piglets were bought at auction, and only after they had been brought home was it discovered that they had atrophic rhinitis. Among other things, this disease attacks the membranes and bones in the nose, eventually curling the snout. In other ways rhinitis seriously inhibits growth.

One of the two pigs sneezed blood periodically, a sure sign of advanced infection and degeneration of the tissues inside the snout. And yet, without any treatment beyond tight housing and a diet that included commercial grower and skimmed milk, the pigs soon stopped their sneezing, all symptoms of the disease disappeared, and they grew on to market weight at an acceptable pace.

SOME COMMON HEALTH PROBLEMS IN SWINE

There is no substitute for local information and vetinary assistance when health problems arise. The table on pages 180 and 181 and accompanying text are meant to provide information about swine diseases and health problems felt to be those most common across North America. Diseases not mentioned may be more important in your area.

A diagnosis based on information provided here may be in error when a locally important disease not covered is characterized by a combination of symptoms similar to those of a disease we have included. Please be aware, and keep in touch with local authorities.

Infectious Diseases

BACTERIAL AND VIRAL

Abcesses: Pigs are prone to bacterial infections that cause abcesses—swellings that eventually fill with pus, like deep boils—at times along the backbone causing lameness or paralysis. They may be under the skin, deep within muscles, or in lymph glands.They may be as small as marbles or as large as baseballs.

Any break in the skin may allow the infecting organisms in. Dirty hypodermic needles, bites, cuts, punctures—all may lead to abcesses. A specific antibiotic may be needed to combat a strain of bacteria infecting a herd or premises. Some books warn against lancing abcesses because the pus may contaminate the environment of the sty.

Atrophic rhinitis: Caused by bacteria attacking the fine network of bones within the snout. AR may appear in young piglets, with sneezing being the first symptom. Pigs rub snouts. Discharges occur from nose and eyes—dark patches appearing beneath eyes from dust collecting on moisture. Coat may appear rough. As infection intensifies pigs may sneeze blood. In advanced stages there may be a noticeable foreshortening and twisting of the snout.

Treatment with specific drugs is not always satisfactory. Heavy culling of sows and infected pigs may help on farms where disease is entrenched. Better environment—dust-free, dry and with lowered concentrations of animals—is encouraged. Although pigs do not die, they may not be thrifty or economic.

Baby pig scours: Several conditions or disease organisms may cause piglets to scour. All are serious because piglets with diarrhea quickly become dehydrated and starved. As any of several diseases may be involved, inexperienced managers should consult a veterinarian whenever there is an incidence of watery, fetid or discolored feces.

Colibacillosis (white scours) is a bacterial disease of suckling pigs. Profuse, yellowish-white diarrhea and a fetid odor are described. Piglets often continue to nurse. Death rate may be high in young (7-8 day) piglets. May be highly contagious. Invasion of the bloodstream causes signs of fever including shivering and raised hair coat, and at times the death of piglets that have not scoured (see *transmissible gastroenteritis*, TGE). Colibacillosis responds to antibiotics given orally, but resistant strains of bacteria may necessitate use of specific types. Prevention through assurance of clean, dry quarters, warmth and adequate milk supply is important.

Other diseases causing scours in piglets include TGE, coliform enteritis (usually a week or two post weaning), swine (hemorrhagic) dysentery, salmonellosis (more often during early fattening period), cholera and early pneumonia. Change in the diet of small pigs may also cause mild, temporary diarrhea (see page 116).

Brucellosis: A bacterial infection caused by swine, bovine or goat strains of the Brucella organism. Symptoms include early abortion, stillbirth and the birth of weak piglets. Boars may show enlargement of one or both testicles; temporary or permanent sterility. Lameness and paralysis of the back legs is also reported in mature hogs.

Diagnostic Key to Common Swine Diseases and Health Problems

Diseases, conditions	Sudden death	Sudden onset	Depression	Weakness	Weight loss	Reduced growth	Low temperature	Fever	Nervousness	Loss of appetite	Thirst	Hairless/hair loss	Rough coat	Bumps/swellings	Blisters/spots/bruises	Abnormal color	Shakes/trembles
Abcesses														X			
Atrophic rhinitis					X								•				
Baby pig scours	•											X					X
Brucellosis																	
Cholera			X		X			X	X	X						X	X
Leptospirosis																	
MMA (mastitis, memitis, agalactia)	•		X					•		•							
Pneumonia								X									
Smedi viral group																	
Swine (hemorrhagic) dysntry	•																
Swine erysifelas "diamond skin disease"	•		X					•		X					•	X	
Swine flu		X		X				X									X
Swine Pox																X	
Transmissable gastroenteritis						X				•							
Vesicular exanthema															X		
Dancing pig disease									•								X
Hypoglycemia				X						•			X			X	
Lice																X	X
Mange															X	X	
Fly, mosquito bites																	
Stomach worms					X	X				•			•				
Round worms					X					•						X	
Calcium, phosphorus, vit. D def.						X							X				
Iodine deficiency												X					
Iron, copper def.	•				X											X	
Zinc def.					X							X			X		
Vit. A def.				X	X												
Vit. B-related def.					X	X				X		X					
Riboflavin					X							X	X				
Niacin					X					X			X				
Pantothenic acid					X					X							
B12					X					X							
Vit. E, selenium	•		X	X													X
Rodenticide (anti-coagulant) poison	•	•			X									•	•	X	
Lead poisoning	•	•						X	X								
Mercury poisoning	•	•		X					X								
Mold poisoning	•	•		X					X							X	
Pitch poisoning	•		X														
Salt poisoning	•	•		X						X	X						
Injuries		X		X													
Heat stroke		X	X						•	X	X						
Sun burn		X														X	
Choke		X															•
Sow in heat									X								
Near death							X										
Cold enviroment							X							X			X

'X' Denotes symptoms more commonly encountered.
'•.' Denotes symptoms occasionally encountered or those specific to one age class.

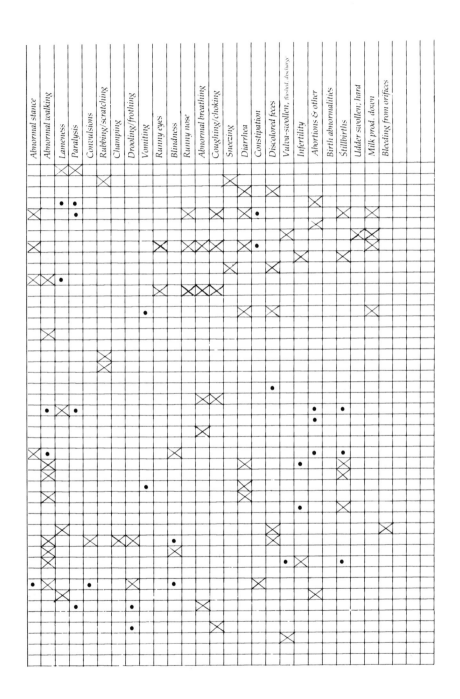

Hogs kept for breeding should be tested for brucellosis. If found to be carriers, they must be destroyed, as there is no treatment. Cases of abortion should be reported so that tests may be conducted to determine the cause—there are many. "Any infectious disease that causes a high fever or severe upset can cause abortion," says *Swine Diseases*, published by Agriculture Canada.

Cholera: The United States has been declared free of hog cholera, but the disease must be remembered and recognized as long as it remains in the world. Cholera is very difficult to diagnose because the symptoms are many and can vary from one case to another. The only symptoms may be unthriftiness, emaciation and occasional coughing and diarrhea. Other symptoms may include nervousness, reluctance to move, trembling, lack of appetite, arched back, flushed belly and ears, temperature of 104°-108°F. (40°-42°C.). Some will show paralysis; inflamed, drippy eyes. There may be constipation early on, followed by diarrhea. Infection often shows up shortly after the introduction of new animals to the herd, and spreads rapidly. Infected pigs usually die within four days of first signs. Canada forbids use of vaccines or attempts to treat cholera. Eradication of disease is the goal through destruction of infected animals. Feeding uncooked garbage containing infected meat scraps is important means of spreading the infection.

Leptospirosis: A bacterial infection causing abortion and the birth of weak or stillborn piglets and lowered milk production. It may be treated with antibiotics. The disease infects other farm mammals, rodents and people. Abortions should always be reported so that a veterinarian can determine cause.

MMA: *Mastitis* (udder infection or inflammation), *metritis* (uterine infection or inflammation), and *agalactia* (lack of milk)—a one, two, three punch affecting sows in the days just prior to and following farrowing—usually within the first three days after farrowing. Symptoms may include a discharge of white to yellow pus from the vulva, inflammation, tenderness and hardening of some sections of the udder, fever, depression, loss of appetite and a drying off of the sow's milk supply. Sows may lie on belly to prevent nursing. They usually survive but piglet mortality can be high.

There are a number of possible causes ranging from hormone imbalance to bacterial infection aggravated by stress and unsanitary conditions. Treatments include antibiotics (oral and injectible) and hormone injections (Oxytocin). Piglets may have to be fostered. Call a vet immediately. Piglets are too vulnerable at this critical early stage.

Pneumonia: Formerly called *virus pneumonia,* now thought to be caused by a mycoplasma (something between a virus and a mold), it is now called *mycoplasma* or *enzooticpneumonia.* The disease spreads rapidly from direct contact among animals, especially sows to their young. A chronic condition begins with a cough often accompanied by diarrhea and a mild fever, 105°F. (40°C.). Coughing is most noticeable when pigs are forced to move about. While other signs may disappear, the cough can persist throughout the life of the pig.

Acute conditions including gasping, "crackling" breathing, wide-legged stance, constipation, drop in milk production, runny eyes and nostrils, can develop with secondary bacterial infection. There is no treatment for the initial, chronic condition. Pigs will do satisfactorily under good management. Antibiotics can be used to treat secondary infection.

Some pneumonia-free herds are obtained by establishing a herd from pigs delivered by Caesarian section into pathogen-free environments.

SMEDI: A combination of viral diseases, symptoms of which are stillbirth, mummification, embryonic death and infertility. "No adequate control measures exist," says *Swine Health,* published by the Alabama Cooperative Extension service. Isolation and cleanliness are emphasized for prevention.

Stillbirth is the birth, at term, of dead piglets. *Mummification* is the birth at or prior to term of fetuses 1 to 2 inches long or larger that are in the process of dry decomposition and resorption. *Embryonic death* may only be noticed as an abnormally small litter, the embryos having been completely absorbed. *Infertility* is a failure of conception.

All of these symptoms can be caused by other problems—be they infections, nutritional imbalances or physical injury.

Swine (hemorrhagic) dysentery: A bacterial infection of the lower intestinal tract of pigs, often in the early fattening period (150 lbs.) and causing severe and discolored diarrhea—yellow-green to blood red or black and occasionally containing shreds of tissue. Sometimes pigs are found dead without having scoured. Those that scour lose condition rapidly, become dehydrated, and can die within two to three days. "Certain antibiotics and organic arsenical preparations are very effective in treating the disease," says Agriculture Canada's *Swine Diseases.* Laboratory tests are necessary to differentiate between swine dysentery and other diseases causing diarrhea.

Swine erysipelas (diamond skin disease): A bacterial disease, usually of young swine, that may cause sudden death without onset of symptoms that include sleepiness, loss of appetite and high temperature. May be a stiff, or "stilted" gait in some, or arched back. On second or third day, light red irregular patches may appear on lower chest and belly, and there may be swelling of nose, ears and limbs.

Diamond skin disease is a milder form of erysipelas with similar early symptoms leading to raised, dark red, diamond-shaped patches that sometimes turn black and gangrenous (sloughing of skin).

A chronic form of erysipelas causing arthritis and enlargement of joints and distortion of limbs, may follow apparent recovery from acute or mild forms of the disease.

Veterinary assistance must be brought in if erysipelas is suspected. Acute infections respond to treatment with antibiotics. Vaccines may prevent infection. Germs are hardy and may live eight months in soil. Diseased animals should be destroyed and carcasses burned and buried with lime. Causes erysipeloid in people.

Swine Flu: Onset of this viral infection is described as "explosive," with symptoms including depression, loss of appetite in some pigs, prostration, difficulty breathing—sit like dogs to assist breathing; very high temperatures—up to 107°F. (41.7°C.), coughing (at times violent when pigs are forced to move about), and discharge from nose and eyes. The disease usually runs its course in three to four days. There is no specific treatment beyond providing dry, comfortable quarters. Deaths are rare unless complicated by secondary bacterial infection. Lungworm is thought to be an intermediate host for the flu virus.

Swine Pox: A viral disease, usually of young pigs. Seen in circular, dime-sized, raised pox lesions appearing first on belly and insides of legs. Lesions become covered with brown scabs. Scabs flake off, bearing virus germs that may last a year or more. Secondary infection may complicate matters for the pigs. As lice may spread swine pox, control of these parasites may help prevent outbreaks. No treatment is available for treatment of the pox.

Transmissable gastroenteritis: A viral disease in piglets causing diarrhea that is whitish, yellowish to greenish in color. Milk may pass through a piglet apparently unchanged. Vomiting and excessive thirst may be noted. There is rapid dehydration and weight loss. In sows there may be no symptoms in some while others lose their appetites, vomit and scour. They may cease giving milk and lose weight.

Mortality is very high in piglets under two weeks of age. Some sows die. Animals that survive carry natural immunity.

Severe scouring as described demands veterinary attention immediately for proper diagnosis and treatment. Antibiotics may help control secondary bacterial infections. Sows in gestation may deliberately be infected by way of contaminated feces to encourage development of antibodies (immune response) that can then be passed on to their piglets in the first milk (colostrum).

Vesicular exanthema: May be transmitted in uncooked garbage containing contaminated pork trimmings. Blisters (vesicles) erupt in the mouth and on the feet. Symptoms are very similar to those of swine vesicular disease and to hoof and mouth disease, and so incidence must be reported to allow for medical diagnosis. No treatment or immunization for VE mentioned.

Dancing pig disease and hypoglycemia: These apparently non-infectious diseases are sometimes confused in the literature available to farmers. In dancing pig the animals show tremors, particularly of the back legs, and twitching to the point that the animals may jump off the ground when they try to move. Noise or excitement aggravate the condition. The cause is unknown, and it is said that if the piglets survive they usually recover fully.

In hypoglycemia, also called *baby pig shakes* or *dancer pig,* the piglets tremble, have rough coat, huddle together and may collapse with legs galloping. They may fail to nurse. One source says the condition is inherited. Others say it is the result of piglet starvation due to agalactia (lack of milk). Treat with heat lamp or hot water bottles, and fostering to a sow with milk, or force-feed a formula at frequent intervals.

EXTERNAL AND INTERNAL PARASITES

Lice: Pigs rubbing vigorously with subsequent loss of patches of hair may indicate presence of lice. Many treatments are available including Rhotenone dust. Treatments should be repeated at 12-day intervals to assure catching lice that were in protected egg stage at time of first administration.

Mange: A skin condition caused by mites, tiny spider-like organisms too small to be seen, that burrow beneath the skin. Symptoms include intense itching, redness and a thickening of the skin. Eventually the skin becomes scaly and brown. Said to be "the only disease that causes pigs to scratch and rub themselves constantly when they are awake." Repeated treatments with available drugs may be necessary to assure eradication. Because other skin diseases may be associated with mange, a treatment for the mites alone may not fully alleviate problems.

Stomach worms: There are three different species, all of them reddish in color and ranging in size from tiny, thread-like worms to those an inch long. Symptoms include poor growth, loss of condition, inflammation of stomach, at times causing the passing of blood in the feces. Appetite may be suppressed or remain normal. Specific drugs are available. Probably should be treated following tests to determine specific worms involved.

Roundworms (ascarids): May be difficult to detect although growth is depressed in mildly infested animals. More severe infestations cause loss of appetite, loss of condition, stunting, coughing, "thumps," or spasmodic breathing in hogs suffering migration of larvae through the lungs. Blockage of the bile duct may cause a yellowing (jaundice) of light skin and mucous tissues. Diarrhea and rough hair coat are also encountered. Worms are whitish and are up to 10 inches long. Roundworms are the most common and costly parasite in swine across North America. Migration of larvae through the liver leaves white scars called "milk spots."

Periodic worming with specific drugs following instructions is required. Also cleanliness and pasture or lot rotation are necessary to break the cycle.

Trichinosis: Clinical signs may not show in pigs, although with a heavy infestation of trichina worms there may be diarrhea. Rats can carry trichinosis and swine can pick the disease up by eating dead, infected rats. Contaminated raw pork scraps are another source for infection. The immature worms encapsulate themselves in muscles of swine, rats and people.

Other worms: Whipworms, kidney worms, lung worms and nodular worms are important parasites in different regions of North America. Look to extension bulletins in your area.

Nutritional Imbalance

Calcium: See *Vitamin D*.

Iodine: Iodine deficiency causes the birth of hairless or nearly hairless piglets, either dead or weak and likely to die. Sows must be provided iodized salt.

Iron and copper deficiency: A deficiency of iron, which may be aggravated by a deficiency of copper, causes nutritional anemia, characterized by poor growth, a rough hair coat and paleness, especially noticeable in mucous membranes (beneath eyelids, gums) of piglets ten days or older. Some piglets have labored, jerky breathing ("thumps"). Occasionally piglets are found dead without having shown any clinical signs of anemia. The condition is most apt to strike the fastest growing piglets. (See Chapter 14 for corrective measures.)

Phosphorus: See *Vitamin D.*

Zinc deficiency: A deficiency of this mineral causes a disease called *parakeratosis,* usually in pigs one to five months old. The signs include pimple-like spots leading to a mangy look (brown- colored, crusty skin), reduced appetite and growth. Parakeratosis appears to be associated with an excess of calcium in the diet.

Vitamin D, calcium and phosphorus: A deficiency of vitamin D or calcium causes rickets in growing pigs, characterized by poor growth, rough coat, lameness and relative enlargement of joints, head, shoulders and hips as the long bones fail to grow. Bones become soft, may fracture easily, and feet may knuckle under. In adult stock these deficiencies can cause bone fractures and posterior paralysis (often the result of a cracked pelvis). Deficiencies of calcium, phosphorus, or an imbalance between the levels of these two minerals in the diet (see Chapter 10) may be responsible for the sows having small litters and for the birth of dead piglets.

Suspected deficiencies can be corrected with injected vitamins and by feeding commercial mineral supplements. Or you may get the pigs outdoors for vitamin D and feed them egg shells, bone meal or wood ashes to appetite. See *Zinc deficiency* for effects of too much calcium in a diet.

Vitamin A, carotene: Night blindness is an early symptom of a vitamin A or carotene deficiency in pigs or hogs, perhaps only noticed if animals are forced to move about in dim light. There may be decreased growth in pigs, followed by uncoordination, posterior weakness, a tilting of the head, and a developing sway-back appearance. Sows deficient in vitamin A or its precursor, carotene, may produce small litters, or their piglets may be born dead, weak, blind, with poorly developed eyes, or with other deformities.

Vitamin A, D and E preparations may be provided for immediate alleviation of the deficiency. Over the long haul a diet including greens, yellow root crops, yellow corn, or well-cured forage crops—or fish oils in young pigs—will provide sufficient vitamin A value. (See Chapter 10 for further notes.)

Vitamin E and selenium: Apparently related, these two deficiencies can cause sudden death without there having been earlier signs, that include weakness, loss of appetite, reluctance to move and trembling. Vitamin E, although present in many fresh foods, can easily be destroyed in storage or processing. Pigs fed rations severely deficient in the mineral selenium seem to require above normal amounts of vitamin E. If deficiencies of vitamin E or selenium are suspected, consult your veterinarian or extension representative. Fresh foods will provide vitamin E, but selenium will have to be introduced artificially into the diet, and too much of this mineral can be toxic.

B vitamins: There are at least eleven compounds making up the B group of vitamins. The four most apt to be found deficient are riboflavin (B-2), niacin and pantothenic acid—most frequently in grain-fed hogs—and B-12, the "animal protein factor," especially important for animals in dry-lot confinement.

The most important deficiency symptoms are listed in the diagnostic table. Further clarification or expansion of the table data is as follows: Riboflavin, abnormal gait in pigs and birth of crippled or deformed young; niacin, occasional vomiting; pantothenic acid, "goose-stepping" gait; and B-12, "lowered reproduction" (Ensminger). Commercial B-complex supplements, alfalfa meal and milk, fish or meat products are usual sources for additional B vitamins.

Poisoning

Whenever pigs quite suddenly become weakened, depressed, stop eating, show signs of nervousness or lost coordination, drool, vomit or have lowered temperatures, and when no other cause is apparent, poisoning should be suspected. Consult a veterinarian.

Anti-coagulant rodent poisons: These poisons, *Warfarin* being one, kill rats and other rodents by interfering with clotting, thereby causing internal hemorrhaging. A pig would normally need more than one exposure to the poison or rodents killed by these chemicals before it would take sick. Effected animals bruise easily, show swellings on limbs, become weakened and lame. A primary symptom would be the appearance of large areas of hemmorrhaging beneath the skin of the belly and inside the legs, looking like large bruise. Sicker animals may pass blood from body openings. Treatment includes injections of Vitamin K to promote clotting.

Lead: Loss of appetite is usually the first sign of lead poisoning. Feces will be dark and may be tinged with blood. Champing of the teeth, drooling, excitement, lack of coordination, blindness, convulsions, coma and death may follow in quick succession. Onset may occur several days after exposure. Sources of lead include old painted surfaces, storage batteries, or plants that have been sprayed with lead-containing compounds. Stamm says a "purgative dose of Epsom salts" is a common treatment for acute lead poisoning. Feed protein (eggs, milk, blood serum), says Ensminger.

Mercury: Loss of appetite, weakness and an unsteady gait, blindness, unconsciousness and death may occur within five to ten days in pigs that have eaten grains treated with mercury-containing fungicides. In other cases pigs may die without having shown any clinical signs. As in lead poisoning, the feeding of protein—eggs, milk or blood serum—is mentioned by Ensminger.

Mold: Acute poisoning from mold (myco) toxins causes loss of appetite, weakness, staggering, and pale or yellow (jaundiced) membranes of the eyes and gums. Death may follow in two or three days in acute cases. Sows may show swollen vulva and anus and may encounter reproductive problems including failure to conceive and the loss of piglets through abortion, stillbirth, etc. Piglets and young gilts also may show flushed, swollen vulvas. Boars show swollen sheath and anus. Straining may cause prolapse (a turning inside-out) of the uterus or anus. (See Chapter 12.) As internal hemorrhaging is involved, vitamin K may be administered. Consult veterinarian.

Pitch poisoning: Depression followed by death is all that is said about pitch poisoning by Ensminger. Other sources don't mention this danger to pigs that eat clay pigeons (trap-shoot targets) or "bitumen-containing products including roofing materials, certain types of tar paper and plumbers' pitch." Ensminger says there is no known treatment. Clean up that pasture!

Salt: Salt poisoning causes many symptoms similar to those brought on by lead poisoning, with the exception of champing and dark or bloody feces. Specifically salt poisoning is said to cause nervousness, constipation, frothing, weakness and a wobbling, staggering gait, and convulsions in severe cases. Pigs may be blinded and may move backwards, sit like dogs, or roll over backwards. Death can occur within a few hours. Salt poisoning symptoms may follow restoration of a watering system as salts accumulated within the body draw water, causing swelling, particularly of brain tissues. Avoid possible over-accumulation of salts by providing ample fresh water. Don't overlook need for fresh water for swill or whey-fed pigs. Where salt accumulation is feared, restrict water intake for three or four days before returning to free-access.

Injuries

Lameness: In addition to the many diseases or nutritional imbalances that may cause apparent lameness, poor floors or injuries such as sprains or cuts or punctures may cause pigs to limp or to favor one or more feet.

Heat stroke: Restlessness, excessive thirst, labored breathing, at times with mouth hanging open, salivating and elevated temperature may be signs of heat stroke, especially when seen in mature hogs on hot days. Paralysis, coma and death may follow. Provide more ventilation and sprinkle animals with cool (not ice-cold) water.

Sunburn: Light-skinned pigs are more susceptible to sunburn. Ears, head and back appear reddened. Sows will fail to accept the boar. Skin may peel off in a thick crust. Coat burned skin with lard or baby oil.

Choking: Pigs wolfing down foods may choke. If something like an apple gets lodged in the throat, death could result. Signs of choke may be coughing, slavering, and in severe cases, a marked paleness may be noticed in light-skinned animals.

Clapping the pig on the chest may help dislodge the obstruction. As a last resort the pig may be restrained with a snare, the mouth held open with a forked stick, and a probe of plastic hose used with care to try to push the obstruction on into the pig's stomach. This, of course, is an emergency measure when the pig may die before a veterinarian could be called.

Fever: Any bacterial infection, or heat stroke, can cause an elevation in temperature. The infection may be secondary to whatever disease or conditon initially lowered the animal's resistance.

Subnormal temperature: A subnormal temperature may be encountered whenever an animal has become so sick it is near death.

Reproduction Failures

In boars: Hormone imbalance, overwork, chronic arthritis, physical injuries, rough treatment from the handler or a sow, and just plain low sexual urge may at one time or another lead to boar failures either to mate with or successfully breed sows. Professional help should be obtained. Hot weather also may cause a temporary loss of desire to breed.

In sows: Any apparent inabiility to conceive is classed as Infertility in a sow. She may fail because she did not develop fully or properly, because of hormonal disruptions perhaps caused by "cystic ovaries," or because of management errors. Gilts crowded together may not be psychologically stimulated to come into heat, or be affected by infectious disease or nutritional difficulties. Consult a veterinarian when problems develop. It does not pay to keep a sow for months hoping she will come around.

Some Books on Swine Health

Swine Diseases, Publication 1484. Agriculture Canada. Information Division, Canada Department of Agriculture, Ottawa, K1A OC7.

Veterinary Guide for Farmers. G.W. Stamm. Hawthorne Books.

Keeping Livestock Healthy. N. Bruce Haynes. Garden Way Publishing, 1978.

Diseases of Swine. 3rd edition. Edited by H.W. Dunne. Iowa State University Press.

Livestock Veterinary Health Service. Dan W. Scheid. The Highsmith Co., Inc.

CHAPTER 16

Slaughtering

Slaughtering in this chapter covers two things: killing and early preparation of the carcass. The killing takes no time and little effort, but it is one of the most difficult chores in raising pork. Preparation of the carcass does take time and effort, but really it is only another job, and not a bad one given the tools, a proper place to work, and a bit of experience. Experience should be gained first with smaller animals such as chickens, rabbits or lambs, where size does not compound problems you may run into.

Two aspects of butchering are *most* upsetting to people slaughtering their first pig. One is the frightful shrieking you may not be able to avoid. The other is the strange look to everything inside a pig. I hope the illustrations included here will help on that score.

The time for killing on a small scale is the fall of the year. Leftovers from the harvest plus enough grain to fatten have been run through the pig, and it is a season when nights are cool enough to chill a carcass left hanging in a tree or a barn. With more luck still there will have been a stiff enough frost to have blasted the fly population.

Even in the U.S. South, fall is the traditional time for slaughtering on-the-farm porkers. "In days-gone-by hogs were slaughtered during the first norther or when nights were cool enough to chill the carcass," writes William Thomas, swine specialist at Texas A & M. "In south Texas this usually was around December 1, while in north Texas people were able to slaughter around November 1."

Today, Mr. Thomas says, most producers send their pigs to

slaughter houses or locker plants. Lockers are popular in Vermont, too, and at the prices I have heard they charge for slaughtering I am not surprised. If a professional offers to kill and cut up a pig, make sausage and package the works for something in the neighborhood of $20, consider it seriously before going ahead with what I have to offer here.

It is probably best not to kill a female pig that is in heat. Her elevated temperature will slow carcass chilling. Some say there is a boar-like taint to the meat. I have not been able to find any scientific backing for either idea, but all asked, from packing house owners to swine specialists, have said that, given a choice, they would wait until a heat was past.

The basic steps in slaughtering are killing, bleeding, cleaning (scraping, skinning, gutting) and chilling. The reason for killing is obvious. The reason for bleeding is to speed chilling of the meat— the sooner it chills the sooner spoiling bacteria are discouraged. Too, with pork, the more easily and completely will the pickle or salt move into the hams and bacons if they are free of blood.

We clean pigs—outside and in—for sanitation, again to hasten chilling, and to remove parts that are not palatable or that are inedible. Most animals are cleaned outside by skinning, and some people do skin pigs, although the job is harder than with other beasts because there is no clear separation between pork fat and hide. You can skin a rabbit as easily as you skin a banana, but a pig is more like trying to skin a watermelon.

A fully cured bacon or ham keeps better if the skin is on, but this is not important to people who lightly cure and then freeze their cuts. Neither would it matter with old hogs that may be too large or otherwise unsuited for curing. We have skinned pigs, but I have not liked the job or the hacked-up product. This is not to argue that scalding and scraping is best. There are different ways to carry out every step in hog slaughtering. Arguing over "best" ways is silly, because what is best is what works with a person's skills, tools and place.

As we go along here I will mention as many alternate methods as possible, but the emphasis will be on a system that takes a minimum of equipment and that I think should suit most people who are slaughtering only one or two animals a year.

Preparation

EARLY PREPARATION—THE DAY BEFORE

Withhold food *but not water* at least 18 hours prior to the time of killing. This is for sanitation. An animal usually defecates when it is killed. An empty bowel minimizes the feces that get under foot or into the scalding tub. Also, an empty bowel is less apt to get punctured in the course of gutting. A third advantage is a hungry pig that may more easily be led to slaughter with a dish of food.

Gather and prepare tools—cleaning, sharpening and so forth. A list of tools and equipment includes:

1. A barrel or tub for scalding. A steel drum such as we use should be fitted with handles just above center through which poles can be inserted for easy lifting by two people. A tub into which the entire pig may be rolled on ropes or chains would probably be a good investment for people doing a number of pigs. (See illustration, page 200)

2. A thermometer that registers up to 200°F. (Dairy thermometers are good.)

3. Pulleys or blocks. (See also item 17.) You may have the people power on hand to lift a carcass (or a tractor's front-end loader) and not need blocks. Or perhaps a single pulley will do. But we find a single and double block make it possible for one person to hold a carcass in the air while others shift tables, move barrels of water, or whatever.

4. Rope. Under most circumstances you will want 20 or more feet of a quarter-inch synthetic or half-inch fiber rope.

5. A rifle .22 caliber or larger. If you use a .22, fire long rifle cartridges. Although the .22 rifle is the popular choice for stunning hogs I have come to favor something larger for those of us who kill but one or two pigs a year. The pig's brain is

A front-end loader makes lifting easy.

small and the skull around it is thick. An experienced person can expect better than 90 percent success stunning with a .22. But I have seen the best miss their mark on the first shot.

After using a .22 for several years with mixed results I have killed the last two by shooting them behind the ear, in one case with a .410 shotgun (firing a slug) and in the other with a .32 rifle. In both cases the stunning was so instantaneous I will probably not go back to .22s. There is no massive destruction of meat or creation of blood clots when larger weapons are used. The only possible drawback is the added cost of larger shells. A .22 pistol is not good, especially when shooting a pig in the forehead. It is not powerful enough.

6. Some people like to have a sledge hammer or single-bladed (pole) axe on hand either for initial stunning or as an emergency back-up to gun stunning.

7. Two knives. One short and stout for sticking—say with a six or eight-inch blade. A hunting knife will do. The second knife may be longer, and will be of greatest help in cutting up the chilled carcass.

8. Two or more scrapers. Knives can be used for scraping but "bell" scrapers are an outstanding invention for this job.

9. At least one meat hook. A hay bale hook will do.

10. A foot or two of tough string to tie off the bung (rectum).

11. One gambrel stick or single tree for each pig—unless you follow the procedure suggested under item 17.

12. Tub or wheelbarrow for the entrails.

13. Clean pots for the liver, heart, leaf fat and small intestines. Perhaps another for the head unless this is being thrown away. A four-quart pot for the blood.

14. A meat saw. If you haven't one of these, a carpenter's saw will do or you can use a cleaver, hatchet or axe to split the carcass in half down the backbone.

15. A prepared location, sheltered but away from buildings, for the fire and scalding tank. Gather firewood, figuring you need

Make sure you have plenty of firewood on hand.

at least three wheelbarrows of good wood. Having too much is a lot better than running short in the middle of doing a dozen other things.

16. A clean barn floor beneath a high beam or some other protected place (beneath a tree limb or where a tripod of 15-foot poles can be raised) for lifting the pig.

17. Optional: A tripod or ladder propped against a building. I have heard about raising a pig this way and it sounds reasonable. Lifting to chill the carcass could be the only time the pig is raised if it were scalded and scraped on the ground, using water dipped from a barrel over the fire. Gambrel sticks would not be needed either if back legs were tied to opposite ends of a ladder rung or strung over branch stubs on the tripod to keep the carcass spread.

Hanging pig carcass without block & tackle.

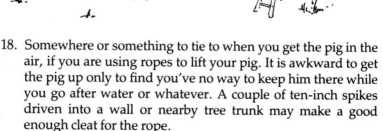

18. Somewhere or something to tie to when you get the pig in the air, if you are using ropes to lift your pig. It is awkward to get the pig up only to find you've no way to keep him there while you go after water or whatever. A couple of ten-inch spikes driven into a wall or nearby tree trunk may make a good enough cleat for the rope.

19. A work surface up off the ground. Grab an old door and throw it across a couple of saw horses. It's a great help.

20. Sawdust to keep things safer and cleaner underfoot, especially when working inside.

21. A hose and tap for fresh water to flush down the carcass—a
helpful extra.

SLAUGHTERING DAY—WATER FOR SCALDING

If you plan to scald and scrape your pig(s), get a fire going beneath
about 20 gallons of clean water, two or three hours before you plan
to kill. Many use wood for fuel. Some use old tires cut up, but
though they make a hot fire, burning tires stink and lay a lot of oily
soot around.

Missouri scalding tank.

The water must be 150° to 160°F. (65°C.), and even a few degrees
hotter on a very cold day or if you plan to ladle water over the pig
rather than dip him in. But do not let the water get above 170°F.
(75°C.), or you risk cooking the skin. This shrinks the hair follicles
and "sets" bristles so that they canot be pulled out with scrapers.

It is said the water is the right temperature if you can stand to dip
your finger in twice but not three times in rapid succession—but one
person may be more sensitive to hot water than another. Then you
hear that a slug of blood thrown in the water will turn white if the
water is hot enough. But so will it turn white if the water is too hot.
Others say to remove the tail and dip that for a test. None of these
matches a $3 thermometer for those of us who kill pigs on occasion.
Stir the water before taking your reading.

Some people add lye, wood ashes or baking soda to the water as
ways to an improved scald and/or a cleaner carcass. I've also heard
of the freshly scalded pig being sprinkled with resin, which melts,
adheres to the bristles and allows for the whole to be more or less
rubbed off without scraping. I don't believe any of these is necessary

Poles across light frame keep the barrel above the floor of the fire pit.

when the water is the correct temperature. This is the key to a good job.

PREPARATION FOR KILLING

Get the pig to the spot where the killing will be done with as littl fuss as possible. An excited pig flushes, with blood rushing to skil and surface muscles. Bleeding will be more difficult and likely not a thorough as with a pig kept calm. Don't slam the pig around. Bruises don't bleed out.

Sometimes a pig is killed and stuck in its pen and then dragged to where it will be scraped or skinned. You definitely lose the blood this way. Also, if the pig is dragged about before the carcass has bled out, there can be bruising.

A killing crate with a hinged side that drops down after the animal is stunned is a big help, since it closely confines the animal while you are lining up a shot.

Unless the pig is confined in a chute or crate, a rope on a hind leg will keep it from running away. This is especially helpful in the unhappy event the pig is not adequately stunned with the first shot or blow. (Some people kill a pig by hanging it by a hind leg, followed by live-sticking. There is no chance for a wounded pig to run off this way, which in my mind is the only positive thing to be said for this slow-kill method.)

The typical way to get a pig to stand still for stunning is to offer it a dish of food or a few apples.

Slaughtering

KILLING

Shooting and live-sticking are the two most common methods for home killing. Many packing houses use electricity to stun. Although I feel less and less easy around a loaded rifle I prefer putting a pig down quickly to having it pump its life slowly away through a gash in its throat. For those who dislike firearms and live-sticking there is the third option—stunning with a blow to the forehead with a sledge or the back of an axe.

Some who favor live sticking say that this is the only way to get a thorough bleed. The idea is that stunning stops the heart, but this is not true. No matter what happens to the head, the heart continues beating long enough to empty the system.

Pigs may be shot in the forehead at a spot just above and between the eyes. On some pigs there is a slight but noticeable hollow there. Or you can draw imaginary lines from the base of the ears to opposite eyes and shoot at the point where they cross.

Where to shoot.

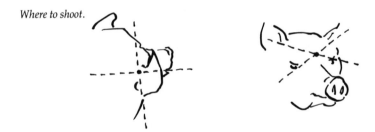

The barrel of the rifle must be held at a right angle to the line of the snout and head or else the slug may travel above or below the critical point—which is through bones, sinuses and into the center of the brain. When you are standing in front of, and over a feeding pig it is easy to shoot low, missing the brain altogether.

The other chosen target for many who shoot pigs is behind the ear, with the shot being aimed down and toward the snout.

When stunning is right, a pig drops like a sack of feed. But it may lie still only a moment before a period of spastic kicking sets in.

Although undirected, a wayward trotter between the legs is no less painful. Be especially careful to keep your face away from those powerful back legs.

STICKING

In sticking a pig you may work with the animal rolled on its back or with it raised in the air. On the ground one person usually straddles the pig (see Chapter 7 for the way to throw a live pig) holding its front legs up, while another wields the knife. With more experience one person may be able to hold and stick. Roll the pig on its back and hold it there with one hand on the off trotter and a knee against its chest while the free hand does the cutting.

Many prefer to lift a pig before sticking. Loop a rope or chain and ring around a hind leg *between the dewclaws and hock* (to avoid bruising the shank end of the ham)—it could be the same rope you already have on the pig to keep him from running away.

Up or down, the object in sticking is to work like a surgeon inserting a knife to sever a major blood vessel, the carotid artery, without hacking up muscle. A great slice from ear to ear is *not* the way to do it, because you risk cutting into shoulder meat. Also this leaves the jowl meat (which is worth dollars) as part of the head, which is worth much less per pound.

The artery you are after lies about three inches below the surface of the skin beneath the point of the breast bone. You can see and feel this bone between the shoulders. Practice finding it before slaughtering day. The carotid artery runs diagonally across the neck at this point so that a neat little cut up the midline of the throat does the job.

Place the tip of a sharp knife, edge down, against the center of the throat just in front of the breast bone. The point should be toward the base of the pig's tail. Drive the point straight down to the backbone. Now pull toward the chin while at the same time bringing the blade perpendicular to the pig's body. The cut should be three or four inches long and, if it has been done right, blood should gush forth in a heavy, pulsing flood. Sometimes the way the pig lies will tend to close the wound and give the impression of a bad stick. Open it with your fingers or roll the pig's head to one side. If the blood still does not gush forth, cut again, reaching a tiny bit farther back toward the chest with the tip of the knife.

Hogs may be "stuck" while held lying on their backs or while held in the air by a back leg.

By no means should you drive so far back that you nick the heart. This definitely interferes with bleeding. Also don't cut to either side of the midline because you will get into shoulder and neck muscles, ending up with clots and unbled portions of meat.

Two advantages to sticking a hanging pig are that the chest, neck and head automatically hang in a straight line making it easier to avoid a shoulder stick, and that it is easier to catch all of the blood for a pudding or whatever. A bit more caution has to be taken to avoid sticking the heart, which will be pressed downward in the chest cavity by its own weight and that of the innards above.

A pig stuck while it is lying on the ground should be lifted immediately for proper bleeding. (See above for method.) The pig should be allowed to hang a few minutes until the flow of blood has dwindled to a dripping. If you are saving the blood, place a large pot beneath the pig. Stir constantly with a wooden spoon to minimize clotting, and remove clots that do form.

CLEANING

Whether the pig is to be skinned or scraped, it is a good idea to give the bled carcass a quick hosing or slosh-down to wash off caked blood or mud that might either contaminate the meat or give you an uneven scald.

SKINNING

As noted earlier in this chapter, we prefer scalding and scraping to skinning. However, for anyone wanting to skin, here are some pointers:

A while ago a writer in *Countryside* magazine suggested skinning a pig by hauling its hide off in long, ribbon-like strips following skin-deep, top-to-bottom cuts made with a linoleum knife.

A Missouri bulletin referring to skinning as a way to deal with old sows or stags, says to first cut off all the feet (at the knees in front, but three inches below the hocks in back to insure heel tendons don't pull loose when the hog is lifted). Skinning is begun with the hog lying on the ground and propped on its back between lengths of 4 × 4 timbers.

The belly, back legs and insides of the forelegs are skinned first, beginning at cuts through the skin running down the midline of the belly and out the insides of the limbs. These beginning cuts are made with the blade of the knife held "slightly flat" to avoid cutting too deeply.

Once belly and back legs have been uncovered, the carcass is raised on a singletree or gambrel fastened by loops of rope to the rear tendons. The odd thing about their system is that the hide is left attached to the backs of the hams while skining proceeds around the sides and down the back. They say this helps in doing a smooth job. When sides and back are done, the hams are skinned, then the shoulders, neck and head, with the ears going with the skin.

James Long writes about a slaughterhouse in Germany where skinning was made easier by placing stuck pigs in a closet-like oven long enough for the heat to "lift" the hide. Again from Long come accounts from Europe where people sometimes burn bristles off pigs by touching off piles of straw thrown on each side in turn.

SCALDING AND SCRAPING

Scalding and scraping (instead of skinning) is easy and leaves an attractive carcass. It does not take long for a couple of people with knives and a bell scraper or two—and a tub of water at the right temperature.

Two people can manage a barrel two thirds full of water (left). The handles for the poles on this barrel are a bit too high. They should be centered just above the middle so that a short helper doesn't have to walk on tip-toes. The right temperature (right) is critical for a proper scald and a thermometer is by far the best way to read what you have.

After the pig has bled out we bring on the barrel of hot water, placing it directly beneath the hanging carcass. The carcass is lowered as far as it will go—only about half a pig will fit in a barrel at one dip—but kept in motion most of the time; lifting, lowering and turning to get an even scald. A pig lowered and left still may have a leg cocked against its side or have its belly wrinkled up so that patches of skin are shielded from the hot water.

We dunk, lift, dunk, lift. Maybe we dunk and let rest a half minute, and lift again, then see if we can pinch off bunches of bristles. As soon as the bristles come easily we lift the pig high, take the barrel back to the fire (putting a couple of sticks of wood on to keep the heat) and slide a table into place. The pig is lowered to the table and we fall to scraping. Some people scrape a pig hanging, but I like having it solidly down on a table.

If you are saving feet to pickle or for soups, right when the pig comes out of the scald is said to be the time to remove dewclaw caps and hard hoof coverings by pinching them off between your thumb and the back of a knife. It's much easier said than done, I find. It would be easier if the feet were parboiled before attempting to remove these horny parts.

Always begin scraping at the extremities—head and feet—since these parts tend to cool and dry faster and anyway are more difficult to scrape than the wide expanses of back and belly.

Generally you scrape with the lie of the bristles. Knives should not be too sharp or they will tend to hack into the skin. Bell scrapers hack less, don't jab friends working next to you, and in so many ways make scraping a cinch. They should be a priority on the shopping list of anyone anticipating getting into this business. A wire brush may be useful in cleaning the bristles from the "elbows" and around the trotters and face.

Sometimes by the time we get to the back, it has dried off, and a bucket of water from the scalding barrel poured over it speeds things up again. Others use burlap sacks sodden with scalding water to hold the heat and moisture in.

It is easier to scrape a hog that lies on a table (left). Bell scrapers make scraping easy. Knives (right) are slower than bell scrapers and more easily hack into the skin.

A meat or hay hook driven through the chin from the inside of the mouth provides a means for lifting the pig by the head in order to dip the hind end for scalding.

As soon as the front end is scraped we drive a meat or bale hook through the chin from the inside of the mouth beneath the tongue and tie this to the block and tackle. You could lift the front by means of a hook under the tendons of a front leg, but we use the chin.

We lift the pig, set the table aside and bring the barrel back for dunking the hind end of the carcass. First, make sure that the water is the right temperature. If it is too hot add a bucket or two of cold water. If it is too cool build the fire up and take a rest. Less time will be lost than if you try to work with water that is not hot enough to give a proper scald.

Scrape the back as you did the front, feet first. Cut off the tail if it has not already been removed.

Once the back half is done, we pick the pig up once more and wash both it and the table down with clean water. We find spots we have missed, yet there will be stray bristles we miss at this point. But strays can easily be got rid of the next day either by singing with a propane torch or by shaving with a knife or razor. So you have a bit of bristle stub? Perfection comes with practice.

GUTTING

Now for gutting the pig—taking out all of the internal organs but the kidneys, which, more by custom in home slaughtering, are left hanging against the back wall of the cut cavity. At one time the kidneys were left as a guarantee of a disease-free carcass. If they were spotted or off-color the carcass might be condemned.

The particular caution in gutting is not to puncture the intestines, because what's inside can taint the meat. If you have a spill, wash the meat off with cool, clear water as soon as possible.

Experienced butchers likely will have the pig in the air when they "gut," but if you're not experienced I think it will be easier to begin with the pig on its back. That way the insides lie away from the opening you are making instead of rushing to get out the door. If you haven't help to keep the pig rolled up on its back, prop it up with a couple of poles or 4 × 4's along its sides.

After scoring the belly (left), the "pizzle" on a barrow or boar has to be lifted up from the body cavity (right) with a little help from a sharp knife to tease away connecting tissue.

For most of us the first step in gutting is to mark a route from throat to bung, scoring the skin lightly with a knife down the center of the chest and belly. The penis and sheath, or "pizzle" on a male presents a bit of a problem. The penis lies just beneath the skin of the belly running from the tip of the sheath to just below the anus where the vessels disappear into the gut cavity. To get the whole business out of your way score deeply enough, from anus to sheath-tip, to reach the penis. At the sheath tip make a cut around so that the pizzle can be lifted up and away from the body. Continue pulling and the penis should tear easily from the body wall. A bit of teasing with the knife will part connecting tissues that don't want to give way. Don't bother to cut the penis off. Leave it hanging back through the crotch.

Now to split the front arch of the pelvis or aitch bone: Cut down between the hams, keeping as precisely between them as possible so that the muscle faces are left intact. You will strike a bridge of bone about as broad as two fingers that completes the ring of pelvis around the rectum and vessels leading to the penis. The bridge is almost solid in an old hog and has to be sawed in two, but on a yearling pig there is a weak seam at the top that can be popped apart with the tip of a stout blade. Place the tip against the top of the aitch

Cutting around the "bung" after the pig has been hung (left). It is easier to cut and tie off the "bung" while the pig is lying down, for then there is no tension from the insides pulling the "bung" down. (Right) Tie the "bung" with stout string.

Although the ribs on a market pig can be parted from the breastbone with a knife, many will find it easier to saw through the center of the breastbone. On older hogs a saw has to be used.

(you may not see it but it can easily be found with your fingers) and gently hammer the butt of the handle with your palm. When the aitch cracks the hams fall away from each other.

(Some recommend splitting the aitch after having inserted the gambrel stick, but you may find that the hams on a pig of typical market weight will not easily spread far enough to take the stick until after the bone is parted.)

In order to avoid feces spilling on the carcass, the bung is now cut away from the hams and tied off with string. You can cut as much as a half inch out from the bung without interfering with the hams, so cut wide rather than close.

Now move to the opposite end of the pig and open the chest. We do this in two steps. First cut through the flesh of the neck and chest, down to the bone. Then come up with a knife, deep, parting the ribs along one side of the breastbone—or you may cleave the breastbone and ribs with a saw. A saw is only necessary on an older, mature hog, whose ribs and breastbone are solid and fused. However, I am sure many will find it easier to use the saw rather than the knife method on pigs of all ages.

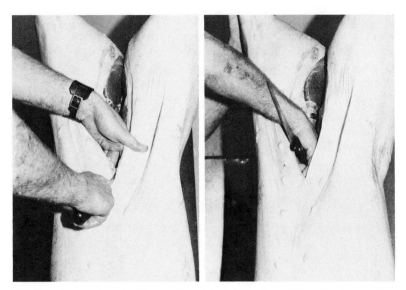

Two ways to open the gut without cutting into the intestines. At left, the fingers of the left hand shield the tip of the knife from the thin-walled bowels. At right, a technique more popular with professional butchers.

If you use a knife, hold it perpendicular to the pig's backbone as you come up the ribs so that the point does not jab into the intestines and stomach, which lie just beyond the thin sheet of muscles (diaphragm) dividing the chest from the lower gut. This knife method takes both hands on the handle and a combination of cutting and prying/twisting motions to part the ribs from the breastbone.

The cut that opens the belly may come up from the ribs or down from the bung. There are two or three ways to cut so as not to puncture the bowels. With the best of these, the knife edge faces you rather than the gut.

One way is to hold and guide the blade between the fingers of your left hand (or right, if you are left-handed) palm up. That way finger tips and knuckles keep the entrails away from the blade.

Another way, one that is more often recommended by professionals, calls for holding the knife with the blade back, edge out, nearly parallel to your forearm. Held this way, the fist and knife handle lead the cutting edge and push intestines and other organs from its path. It is a method that works best with a hanging pig, starting at the opening made for splitting the aitch.

If your pig is not already in the air, now is the time to place the gambrel stick beneath the tendons of both hind legs and hoist him up. To lift these tendons—two of which run up the back of each hind leg between the dewclaws and hocks—make a deep cut about four inches long up the back of the leg. Pry the tendons up with your finger or a hook. Make sure you get ahold of both tendons, especially on heavy hogs. One may not be strong enough. As you lift the pig, the viscera will fall down and forward, but they won't drop completely out of the body because they are still tied in by tissues along the back wall.

Reach in and detach the liver, which is tied into the rest of the system at one point through blood vessels and the bile duct running to the small intestines just below the stomach. As soon as the liver is out remove the small green bile sack (gall bladder) by first severing tubes leading from it two or three inches from the sack itself. Lift

To hang the pig, slice through the skin of the back leg (left) to release the heel tendons. There are two tendons on each leg. One will do for lifting a light hog, but it is better to insert the gambrel stick beneath both tendons (right).

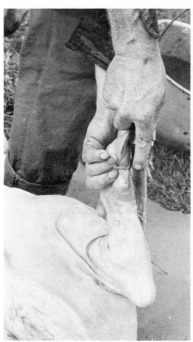

Inserting the gambrel stick.

these tubes at their severed ends and peel back toward the sack. As in lifting the penis it may be necessary to tease a bit with a sharp knife, but only a bit, before the sack comes neatly away. Bile is bitter. Maybe you have tasted it on chicken livers. If you spill bile on the liver or elsewhere wash it off with plenty of fresh cold water.

Now haul the guts out. You may want to bring the tub or wheelbarrow under the pig to catch them. Or perhaps a pile of clean hay or straw may go on the ground beneath the carcass if you plan to remove the small intestines later for chitterlings or sausage casings.

Take hold of the string tying the bung and pull forward through the opened pelvis. Cut the connecting tissues behind. Continue pulling forward and down. Most of the webs of tissue holding the innards back will begin breaking away. The rest will part at little more than a touch of a sharp knife because of the increasing strain of the guts pulling down.

Leave the kidneys and covering leaf lard against the back wall for the time being.

The next barrier will be the diaphragm through which the stomach tube (esophagus) runs on its way to the throat. Cut entirely around the chest cavity removing the diaphragm. Pull the pink,

Pull the "bung" through the split aitch (left). The viscera (right) does not fall out as soon as the gut wall is opened but must be released by cutting and ripping away connecting tissues long the back wall.

Remove the gall bladder as soon as the liver has been exposed (left). The gall bladder (right) comes away cleanly with little cutting.

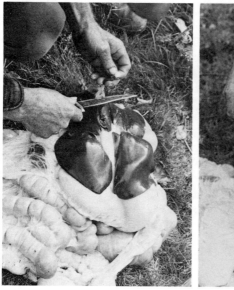 

spongy lungs down to follow the stomach and intestines. Remove blood clots, bits of tissue, blood vessels and whatever else now remains between you and a cleaned-out gut. The leaf lard may be pulled out now. This will help to speed cooling. However, if the weather is cold and you are selling the carcass, there may be more to be gained leaving it in, only with the lower two-thirds lifted away from the carcass wall.

The head may be removed any time after scalding and scraping is done. Out of habit, I guess, we usually wait until the pig has been gutted, but there is no reason not to do it sooner and in fact some prefer to get the head out of the way far earlier in the process.

Starting at the back of the head, just behind the ears, cut to the backbone. Come around each side and follow the line of the jaw to the chin. Once all of the flesh has been cut have someone hold the front feet while you grab the ears and twist the head entirely

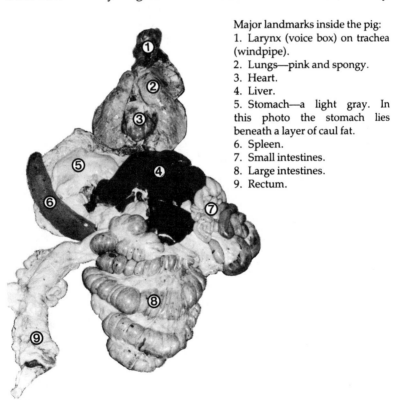

Major landmarks inside the pig:
1. Larynx (voice box) on trachea (windpipe).
2. Lungs—pink and spongy.
3. Heart.
4. Liver.
5. Stomach—a light gray. In this photo the stomach lies beneath a layer of caul fat.
6. Spleen.
7. Small intestines.
8. Large intestines.
9. Rectum.

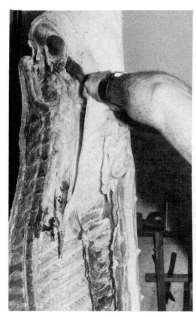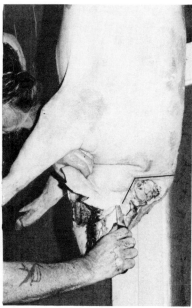

The leaf lard covering the kidneys (left) may be removed to speed chilling. If it is to be sold with the carcass, it may be merely "lifted" away from the gut wall and left hanging from its base just below the hams. (Right) Remove the head carefully, making sure the valuable jowls are left on the carcass.

around. By its own weight the head will fall away, parting from the body at the point where the skull meets the first vertebra.

With perfect aim of the blade cutting between the skull and the first vertebra you may be able to remove the head without the twisting, but this other method is good enough.

The front trotters may be removed at the first joint with a saw.

The carcass now should be given a final wash-down, inside and out, with cold water. Then it may be split in two down as far as the neck, or entirely in two if that suits you better—halves being easier to move than the whole carcass. If you intend to cut the carcass completely in half make sure you have check ropes on the gambrels, "C" clamps, or some other guarantee that the halves won't slide off the stick in the event they are not exactly the same in weight.

A meat saw is by far the best tool for splitting a carcass, though we have used a cross-cut carpenter's saw, and I have heard of going down the center or one side of the backbone with an axe or cleaver.

Saw the carcass in half from the inside. Make long, sweeping passes with the saw. Short, frantic jabs cause the blade to run off to the sides. Keeping the angle on the saw steep also helps to stay on course.

Sometimes it is convenient to quarter the hog—maybe you are selling front or hind quarters—and this can also be done with a warm carcass. The quarter cut is commonly made between the second and third rib, counting from the tail end, but there is no hard rule in this. Only figure that the more forward you go with this cut the more you are cutting through the bacon, which may be of prime interest to the person buying a front quarter.

You can more easily find the ribs inside the carcass, so start here. If the side is hanging, have someone ready to hold and catch the front quarter as it is parted off.

Jab a knife all the way through the side against the backbone and between the ribs you have chosen. Now cut toward the belly, following the path of least resistance between the rib bones. Continue the line of this cut beyond the ribs and to the belly edge of the half. Saw through the back and backbone from the outside to meet the point where your knife cut was begun.

CHILLING

The carcass should hang overnight to chill. Hang it high beyond the reach of cats or dogs. If flies are bad, you should consider building a tent or cage of screening around the meat. But don't wrap the body tightly with gauze because this will only keep the body heat in.

If the weather is on the warm side make sure the carcass hangs in shade and preferably where a breeze is blowing. If there is no breeze maybe you can make one with a fan. For meats destined for curing, Morton Salt recommends an iced brine method for chilling in warm weather. The halves are cut in large pieces and are put down in a barrel of water with chunks of ice and three pounds of salt. They also say the halves may be laid on beds of chipped ice and covered with more. A drier method suggested in Spencer's book is to hang the carcass in a small cellar cooled by blocks of ice placed high, near the ceiling. It takes a lot of ice, though. One sixth of the area of the cellar.

There should be no danger of freezing the carcass unless it is hanging out in very cold weather—say the low 20°sF. (-5°C.) or below. If you do fear the carcass is going to freeze, either find a better spot to hang it, hang a light bulb beneath a shroud over the meat, or start cutting it up when the flesh gets down to the mid-30s F., (2°C.) all the way through. Meat that is too warm or too cold will not cure properly, so this information regarding chilling and freezing may only concern your hams and bacons. You will likely want the entire carcass cooled, however, for the sake of ease in cutting it up in smaller pieces. A warm carcass is a devil to handle, especially when it comes to the smaller cuts such as the chops and individual roasts.

If you are saving the small intestines, cut them off just below the stomach and again where they meet the large bowel. Pull them through your fingers to expel the contents. Turn an inch or two cuff in one end and hold under running water. The weight of the water filling the cuff turns the intestines inside out, exposing a layer of mucous that can be scraped off with the back of a knife. Place cleaned intestines in the refrigerator.

What you do with the head depends on custom and appetite. The best use of the whole head, less the eyes, is in the making of head cheese. But this takes a more thorough scraping than we usually deliver. We skin the head and discard the skin, ears and rooter. Others, foregoing head cheese, may save only the tongue.

Bury the entrails and other wastes deep in the garden.

Wash hands.

CHAPTER 17

Portioning, Processing and Curing Pork

Those halves of pork hanging in the barn may look formidable the morning after slaughter, as you approach with cleaver, saw and freezer bags determined to hew them into recognizable cuts of meat.

Lower your sights a little. It is not likely that your first chops, roasts or slabs of bacon will look like those you have come to know at the market. Neither is it necessary or always desirable that your cuts resemble store cuts.

Although tradition makes us look for those familiar, named chunks of meat, profit guides many professional butchers' knives in the art of show and deception. Most of us have been caught at least once by "specials" on roasts that turn out to be 90 percent bone. And notice the beveled edge along the fat side of a roast or ham. That's done to make the layer of fat appear thinner than it really is.

Can an amateur do worse? So slice away, remembering the most helpful advice I ever got: No matter how you cut it up, it's all going to the same place. Shape makes no difference.

Depending on your stock of tools, you may want to do all of the cutting yourself or to farm out parts of the job. Again, you may find the operator of a freeze locker plant who will do a good job for a reasonable fee. Watch how it's done as preparation for next year's pig.

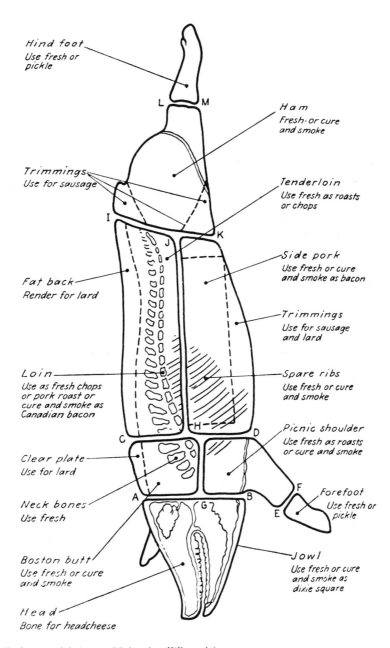

Hind foot
Use fresh or
pickle

Ham
Fresh- or cure
and smoke

Trimmings
Use for sausage

Tenderloin
Use fresh as roasts
or chops

Side pork
Use fresh or cure
and smoke as bacon

Fat back
Render for lard

Trimmings
Use for sausage
and lard

Loin
Use as fresh chops
or pork roast or
cure and smoke as
Canadian bacon

Spare ribs
Use fresh or cure
and smoke

Clear plate
Use for lard

Picnic shoulder
Use fresh as roasts
or cure and smoke

Forefoot
Use fresh or
pickle

Neck bones
Use fresh

Boston butt
Use fresh or cure
and smoke

Jowl
Use fresh or cure
and smoke as
dixie square

Head
Bone for headcheese

Pork cuts and their uses. (University of Wisconsin)

TOOLS YOU'LL NEED

A short *boning knife*, a longer *"carving" knife* and a *meat saw* are just about essential for cutting up your pig. Cleavers may be used, and you might get away with a carpenter's saw, but I think it is worth the ten or so dollars to have a meat saw for those few times a year when it's needed. A wood saw tends to clog and bind.

A *meat grinder* is needed for making sausage. We use a small kitchen grinder, but it is not entirely satisfactory. Large commercial type grinders do a more thorough job of breaking up the meat fibers and mixing them with the fat.

You might want a *sausage stuffer* if you are making link sausage. If so, keep an eye out for one at a junk or antique store. There are dozens of varieties. We have never bothered with link sausage and instead put the sausage up in bulk packages or patties that are immediately frozen.

You do need some *large, clean surfaces* on which to work, not only at the beginning when you are making those first cuts through half carcasses, but later when you have to cut, stack and wrap pieces all at the same time. There is nothing worse than trying to saw when you are hemmed in on all sides with stacks of meat, paper and piles of bones. Clear off a big table and spread out some newspaper. Lay down a clean board for a cutting surface.

WHERE TO LEARN

We learned our meat cutting from neighbors and from *A Complete Guide to Home Meat Curing*, published by Morton Salt. I can't put my neighbors in this book, but the people at Morton Salt have generously granted permission to reprint the portion of their book entitled "Cutting the Pork Carcass." It would be foolish for us to try to duplicate what already is so excellently done.

The Morton Salt book describes curing methods using pre-mixed Morton cures that can't always be bought at your local store. Nonetheless, the information on method is universal and good. The Morton book is now out of print, but you may find a copy at your local used book store.

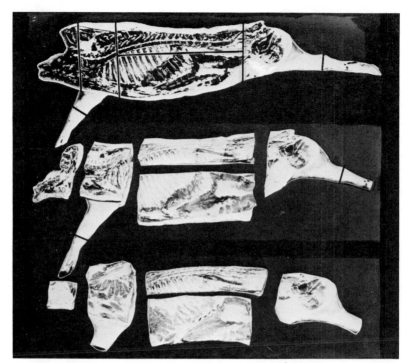

1. *The black guide lines in the top photograph show where the different cuts should be made for cutting up the carcass. The principal cuts are ham, loin, bacon, shoulder and jowl. All remaining pieces can be considered trimmings. By doing a neat job of trimming all of the small extra pieces can be used to greater advantage for sausage, head cheese, scrapple, etc., than if they were left on the larger cuts. Left on larger cuts destined for curing, these trimmings will tend to dry up and will be of little value.*

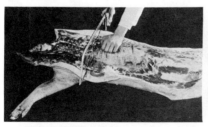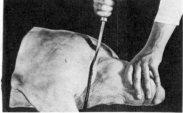

2. *Start cutting up the carcass at the shoulder, sawing through the third and fourth ribs at right angles to the back. Complete the cut with a knife.*

3. *Turn the shoulder over and cut off the jowl at a point where the backbone ends, which is in line with the wrinkle of the neck.*

Meat cutting photos and text adapted from "A Complete Guide to Home Meat Curing," Morton Salt Division of Morton-Norwich.

4. Trim some of the cheek meat from the jowl and flatten it with the broad side of a cleaver or hatchet and square it up by trimming with a knife. The trimmed jowl is known as a ''bacon square'' and can be cured and used the same as bacon or used for seasoning with boiled foods.

5. Remove the neck bone from the shoulder, leaving very little meat on the bone. Trim up the shoulder and cut off the shank. This ''long cut'' method of trimming will give you the maximum of cured meat if this is what you wish to do with the shoulder. Where smaller cuts are desired, the shoulder can be divided between the smallest part of the blade bone, producing a picnic shoulder and shoulder butt. These may be cured or used for fresh roasts.

 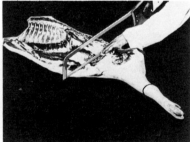

6. When the shoulder is separated into picnic and butt the clear plate, which is the covering of fat on top of the butt, is trimmed off. This may be cured and used for seasoning or may be used for lard. Leaner portions of the plate, called the ''Boston butt'' may be cured or used in sausage. When neatly trimmed, the picnic shoulder has the appearance of a small ham.

7. To take off the ham, saw a line at right angle to the hind shank and at a point about three finger widths in front of the aitch bone. Finish the cut with the knife and start shaping the ham by curving the cut on the belly side.

8. To remove the tail bone slip the knife under the bone and continue the cut along the bone, keeping the knife as flat as possible.

9. After the tail bone is removed, the hams should be smoothed up and all loose pieces of meat trimmed off and put in sausage.

10. To separate the loin from the belly the ribs are sawed across at their greatest curvature. This is about 1/3 the distance from the top of the backbone to the bottom part of the belly edge. Make this cut so as to include the tenderloin with the loin.

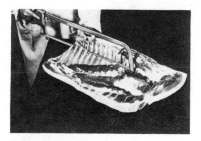

11. After the ribs are sawed through, finish the cut with the knife completely separating the belly side from the loin. Remove fat back from the loin by placing the loin skin-side down; set the knife about one fourth of an inch under the lean or muscle meat, and make a full-length cut. Reverse the loin and make the same cut from the other side. The fat may be cured or rendered into lard.

12. The remaining fat on the loin should be smoothed up, leaving no more than one fourth of an inch on the meat.

13. *The tenderloin, the small lean muscle lying underneath the backbone at the rear of the loin, can be removed with a knife as shown. The tenderloin may be prepared by cutting across into pieces about one and a half inches thick. The pieces may be "Frenched" by placing them on end on a strip of parchment or waxed paper. Fold the paper over the top and strike a sharp blow with the flat side of a cleaver.*

14. *To prepare the bacon, lay the belly on the table skin side up. A few sharp blows with the broad side of a hatchet or cleaver will help loosen the spare ribs on the under side. Now turn the belly over and trim out the ribs. Start this cut by loosening the neck bone at the top of the ribs and keep the knife as flat as possible to avoid gouging the bacon. Pull the ribs upward as the cut is made and trim as close to the ribs as you can. The cartilaginous ends, or "buttons," of the lower ribs are left on the bacon.*

15. *Square up the bacon by trimming the lower edge first to a straight line. All of the "seeds," the mammary glands along the lower edge, should be trimmed out for choice bacon.*

In "Cutting Up the Pork Carcass" there is little on chops, small roasts or fresh ham steaks, because the authors were concerned with producing larger cuts for curing. If you plan to cure only the hams, bacons, jowls and picnic hams, you can cut the loin into chops and roasts, the sizes or thicknesses depending on taste and requirements.

You may like thick chops, even thick enough for stuffing. If so, you can make each one a full rib thick. Cut between the ribs with a knife, all the way to the backbone, and finish off through the bone with a saw or cleaver.

We like smaller chops, and so what we have done in late years has been to place the entire loin in the freezer, let it get solid, and then take it to a store where they can thin-slice it on the band saw. Last year I took the loin and the butt (shoulder) in for sawing into chops and steaks. A band saw also does a speedy job of cutting a ham into steaks, either fresh or cured.

PACKAGING

Pieces of meat for freezer storage should be double-wrapped to prevent drying—called *freezer burn*. An easy method is to use freezer paper for the inside wrap and a plastic bag for the outer. We usually wrap two or more chops in paper, with slips of extra paper between the chops to keep them separate, and then load several of these packages into large plastic bags. Suck the air from the bags before tying them closed.

Label and date packages. Pork does not have as short a freezer life as some charts indicate. But it is always good to use the oldest meats first, to use fatter cuts and sausages within half a year, and to have it all used up within a year. Be sure your freezer stays at 0°F. (-20°C.), or below.

Use the Trimmings

LARD

There will be a pile of trimmings left after butchering. The fat ones may go into a pot along with the leaf lard for rendering over low heat. Or, along with the fat off the loin, they may be cured in pure salt or a sweet pickle for later use with recipes calling for salt pork.

We put fat for rendering on the back of the wood range or over boiling water for the first of it, then move to higher heat so the melted fat can be warmed with constant stirring to drive off the moisture. If you have a frying thermometer take the lard to 255°F. Otherwise go on judgment. Cracklings (bits of skin and meat) will brown and float as rendering nears completion. Then, as the last of the moisture is driven off, they will begin to settle. It's time to remove the lard from the fire. Strain the lard and set it in a cool place. Stirring to aerate the lard after it has cooled to the consistency of cream is said to keep it from getting "grainy."

SALT PORK

Many recipes call for salt pork as one of the added features to a main dish. Diced and fried with onions, it is eaten with salt fish and potatoes. It is often included with lean dishes. But the following recipes call for salt pork streaked with a bit of lean as a dish in itself.

From Long:

"A capital method of rendering salt pork more palatable is to cut it into pieces ready for frying and half fill a crock...with the pieces cut. The whole should then be covered with sweet skim milk, fastened down, and placed in a cool apartment. In six or eight hours it will be ready for use. In frying, the meat may be about half cooked, and then rolled in flour and browned. Two or three eggs and a small quantity of pepper and salt mixed together, in which the meat should be frequently dipped, add to the delicacy of the dish."

And from *Housekeeping in Old Virginia* (John P. Morton & Co., 1879):

"Many people do not relish salt pork fried, but it is quite good to soak it in milk two or three hours, then roll in Indian meal and fry to a light brown. This makes a good dish with mashed turnips, or raw onions cut in vinegar; another way is to soak it overnight in skimmed milk and bake like fresh pork; it is almost as good as fresh roast pork."

Lard should be sealed from air and light and be kept at something less than room temperature. I have some in foil in the refrigerator that has kept well for ten months. Ashbrook's *Butchering, Processing and Preservation of Meat* mentions an old-fashioned way to keep *cooked* fresh meat, which is to pack it in clean crocks covered on all sides with melted lard. The crocks are then left in a cool, dry place. Certainly it is a way to keep unsalted meat over the winter without refrigeration. But more, it demonstrates the keeping qualities of well-rendered lard. Undoubtedly sealed jars would be preferable to my tin foil.

The fat from around the stomach (caul fat) and that from the intestines (the ruffle fat) are rendered the same way, but because they yield a darker lard they are usually rendered apart from the "finer" lard, and the product is used for making soap.

SAUSAGE

Leaner trimmings are the stuff of sausages. They should not be all lean, however. Most recipes call for a ratio of 1/3 fat to 2/3 lean. Cut the pieces into small chunks and run them through a meat grinder twice. Add spices, mix, and grind once more.

A USDA sausage recipe turns up everywhere, and here it is again, since it is the one I have used for a basic mix. It is a simple matter to

---

USDA SAUSAGE RECIPE

To 4 pounds of ground trimmings (1/3 fat to 2/3 lean) add: 5 teaspoons salt, 1 teaspoon sugar, 2 teaspoons ground pepper, 4 teaspoons ground sage, and (optional) one half teaspoon of ground cloves or 1 teaspoon of ground nutmeg.

Mix well and run through the grinder once more. Enough cold water to make the final mixture a bit sticky may help if you intend to make patties or to store the sausage in bulk. It is recommended that no water be added to sausage that is to be stuffed into casings.

vary the spicing as you go along. Have a hot frying pan by your elbow and sample smidgeons of sausage until the concoction suits your taste. If you get too heavy with the spices, as once happened to me when salt and sugar containers got switched, use the sausage in spaghetti or other sauces.

HEAD CHEESE AND SCRAPPLE

Head cheese and scrapple, almost the same products, make use of leaner meats that for one reason or another will not make meals in themselves. The major difference between the two is that more of the cooking broth can be used directly in scrapple because cornmeal or other grain meal mixtures are included. Head cheese uses only enough broth to moisten the minced meats. Natural gelatin gives the cheese body once it has been chilled.

What parts of a pig's head go into the pot depends on personal bias and on the cleanliness of the head at the start. The head should

---

SCRAPPLE

To make a scrapple, pieces of prepared head and whatever else is to be included—feet, heart, tail, etc.—go in a pot and are cooked with enough water to cover. Cook until the meat loosens from the bones and the skin and gristle have become soft. Remove from the broth. Strain and measure the broth. Set aside. Pick the meat, skin and such from the bones and run through a grinder, not too fine. Recombine the ground meats with all but about a cup of the strained broth. This is used to moisten the corn or other cereal meals destined to become part of the final product. This moistening avoids lumps.

About a cup of cereal will be needed to thicken each two cups of broth. After adding the moistened meal to the stew, bring it all to a boil while stirring. As it thickens, add salt, pepper, ground sage, maybe some minced onion, some red pepper—the choices are yours—and continue cooking and stirring until it arrives at the consistency of porridge. Pour into loaf pans and cool. Because of the cereal, slices of scrapple hang together when they are fried. Some people dredge the slices in flour before frying.

be split in half and possible quartered so that it will pack into a large pot. You may want to skin the head because it is still covered with bristles, but do so after splitting because it is a help having the ears for handles while you are sawing the skull. An axe or cleaver can be used to split the head but with a saw there are no bone fragments.

Some people remove the brains. Some remove the teeth—bashing them off with a cleaver. Just about everyone removes eyes, (an easy job once you have quartered the head) and the ear drums.

We often throw away the entire skin, including ears, because they have not been properly scraped, but this is wasteful. With cooking, the skin, ears, rooter and almost all of the gristle in the head becomes soft. At the same time they add nutrients to the broth.

Instead of knocking the teeth out we give them a thorough scrubbing with a wire brush. This is easy if you have first detached the lower jaw.

HEAD CHEESE

The same meats that go in a scrapple can go into a head cheese. Cook as with scrapple. When it is fully cooked, pour through a colander or strainer, and when the meat has cooled enough to handle, pick through it, removing bones and hard gristle. Dice, grind or pick the meat apart. Grind up the skin and soft gristle, rooter and so on.

Return only enough of the broth to the meat to moisten. Add salt, pepper and other spices to taste. These other spices might be sage, red pepper, summer savory, coriander, ground clove or caraway. Bring to a boil, mixing thoroughly. Pour into loaf pans and cool. Once they are down to room temperature the pans may be stacked, with sheets of waxed paper between. Place an additional pan weighted with water or a pound of butter on top. This helps give you a compact product.

Again it is your choice whether or not the tongue is included in the head cheese or scrapple. In this household it usually does go into one of these or into sausage.

Head cheese makes a wonderful sliced cold meat for sandwiches or any cold plate. It may be fried too, but it doesn't hold together as well as does scrapple.

The broth left over from the head cheese may be used to make a form of *ponhaw* that consists of broth and corn meal—sort of a meatless scrapple.

THE PIG COOKBOOK

There may be many, but the only cook book I know of that is devoted entirely to cooking pork is *Charcuterie and French Pork Cookery*, by Jane Grigson, published by Penguin Books.

Curing Hams & Bacon

The curing of hams and bacons was invented before refrigeration, as a way to preserve meats. The majority of commercially "cured" pork today must be refrigerated anyway, because it has only been treated with enough salt and smoke to give it a desired flavor. In fact most doesn't come in direct contact with smoke any more, being instead injected with a mild, smoke-flavored brine.

Most of the preserving action of cures comes about as a result of drying. Moisture is drawn from the tissues of the meat by the salt or salt solution just as salt on a gravel road draws moisture out of the air. Bacteria and other microorganisms of decay find it hard to thrive in the dry climate left behind.

Sugar, molasses or maple syrup may be included in curing mixtures to counter the harsh flavor of salt alone.

Saltpeter is added primarily to assure a pink or reddish color in the cured meat. But because the chemicals—nitrates or nitrites—of saltpeter are not good for us we do not include them in our cures. We were led to expect our bacons and hams would come out shades of deadly grey but they have not. They look fine.

Smoking gives meats a pleasant flavor. Salted meats hanging in a smokehouse or chimney chamber are able to drip and dry even more without worry of being attacked by flies. Long-term (several day) treatments with heavy smoke begins to fill cracks and pores in the meat and so presumably aids in sealing the meat against invasion.

WET AND DRY CURES

There are two basic ways to cure—wet and dry—and sometimes a combination of the two is used. In *dry curing* the salt—and perhaps some sugar as well—is packed over all surfaces of the meat, which then is stacked on a wood or stone table or packed into a crock or a wood or plastic barrel provided with drain holes. Dry curing is considered faster than wet or brine curing, and is more often recommended as the way to cure in warmer situations.

In *wet curing* the meats are packed in a barrel and a brine of dissolved salt and other spices is poured over the top.

A *combination cure calls* for covering the meats with dry salt/sugar, packing them in a barrel, and adding brine to cover.

Wet curing is easier for small batches of meat when it is difficult to get the whole lot packed snugly and level in a barrel. It is also easier for the inexperienced who may find it hard to get dry salt and other spices to stick to the surfaces of the fresh meat.

Meats may be mild or fully cured. Mild cured cuts are more moist and less salty, and haven't the keeping qualities of those that are fully cured. Mild cures usually are obtained by cutting down the curing time.

Recipes for cures vary widely, and yet certain steps are so consistently recommended that they can be looked on as universal rules. They are:

1. With two possible exceptions, always pack meat or fat pork skin-side down during curing. The two exceptions are the top layer in a pack, which may be skin-side up, and, in the case of crock or barrel packing where the skin surfaces may be to the outside.

2. Hams should be packed shank down or skin-side down in a barrel or on a shelf. If they are hung during dry curing they should be hung shank down, and wrapped in a cloth bag if there is any danger of insect attack. When they are later hung for smoking or aging, the hams should be shank up.

3. The curing pork should be unpacked or unstacked and re-arranged at least once, about a week into the curing time, and preferably at weekly intervals thereafter until the meat has fully taken the cure.

4. Thinner cuts such as bacon sides cure more rapidly than do the thicker hams.

5. Don't be afraid of overcuring. To a large extent overly salty meat can be soaked out in fresh water even after smoking. In fact we have "freshened" fully cured and smoked bacons in this way and then returned them to the smokehouse for new smoke. With under curing you risk spoilage and the good chance that everything will have to be thrown away.

6. Brines should be made with pure water, preferably boiled, and then cooled before pouring over the meat. Salt and sugar dissolve more quickly in hot water.

7. Around 40°F. (5°C.) is the ideal temperature for curing. Below 36°F. (2°C.) very little withdrawal of juices takes place, and when the temperature drops below 34°F. (1°C.) you can assume that the curing process has stopped. Above 50°F. (10°C.) you run an increased risk of spoilage. Our curing barrel goes in the basement where the temperature hovers around 50°F. (10°C.) all of the year. We have never had problems with spoilage.

8. Salt corrodes, so stay away from metals: No metal buckets, no wire coat hangers (watch out for nails). Use plastic, stone or stoneware, and wood.

9. Any piece of meat above the brine or not coated with salt can act as a wick, drawing spoilage to the rest of the lot being cured.

10. If at any time during the curing process the meat gets slimy or the brine gets slippery and stringy you know that spoilage organisms are on the increase. Remove and wash the meat thoroughly with fresh water. Filter and boil the brine, or mix a fresh batch and repack the pork. Put fresh cure mixture on meat being dry cured.

RECIPES

HAM AND BACON CURES

Dry curing is probably the oldest and still highly respected means of treating pork for storage. The standard recipe calls for 7 pounds of salt (use coarse, pickling salt and never one of the salts that has been treated to prevent lumping) and 3 pounds of sugar for 100 pounds of meat. This works out to about 1¼ ounces of cure to each pound of pork. Other spices such as black and red pepper, juniper berries, etc., may be added by the teaspoon according to taste.

In dry curing the meat, especially where the flesh is exposed, cover all over with the cure mixture and then place skin-side-down on a sloping table or board in a cool room. A cellar may be ideal. All of the cure may be applied at one time or in two stages, with the second half of the mixture being applied after three days when the lot is overhauled. The cure mixture should be patted on in order to assure the meat is well covered. Two or three layers of meats may be salted and stacked upon one another, perhaps with a bit of extra salt being sprinkled over the meats between layers. Make sure that windows and doors are screened to keep out insects.

Fully (country) cured fresh hams should be allowed to cure for 2½ days per pound, or about 50 days for an average 20-pound ham. Sides of bacon and other thinner cuts may cure in half the time.

I have sometimes had trouble getting the dry cure to stick properly to the meat and find that it helps if I first smear molasses or maple syrup over the surfaces. Some recipes call for heating the dry cure in a pan and applying it to the meat as soon as it is cool enough to handle. This mixture should stick better.

A mild cure offered by the University of Missouri calls for the same 7:1 ratio (by weight) of salt to sugar, with 2 pounds of the mixture being dissolved in a gallon of water. Once cooled, this pickle is injected by syringe in and around the joints and bones of the hams at a rate equivalent to 10 percent of the weight of the cut—approximately 1 quart (2 pounds) to a 20-pound ham. The remainder of the dry cure mixture is rubbed over the meats at a rate of about ½ ounce per pound. The hams are placed on shelves where the temperature remains between 34 and 45 degrees F. for 14 days. Sides of bacon are allowed to cure for seven.

The method we have used most often for mild curing calls for a pickle made by dissolving 7 pounds of rock salt and 3 pounds of sugar (brown or white) in 3 gallons of boiling water. Others mix enough salt to float a medium size potato in enough water to cover the meats packed in a barrel. Varying amounts of sugar (according to taste) are added to the brine. After cooling, the pickle is poured over the meats. The whole is overhauled at least once within the first week and perhaps a second time before the end of the curing period. A plate weighted down with a clean stone keeps the meats from floating to the surface of the pickle. Bacon slabs and other smaller cuts may cure in 10 to 14 days. Hams require five to seven weeks.

To produce a more moist ham, try this method. After taking the ham from the pickle, wrap it in cheese cloth to hold it together and simmer in fresh water for 8 to 10 minutes per pound. The partially cooked ham is then smoked for five days to two weeks—the shorter period if the weather is dry, with temperatures hanging around 50°F., and longer if the weather is cool and damp. Prior to eating the ham should be roasted for 10 to 12 minutes per pound at 375°F.

Methods for protecting smoked hams and bacons from insects include wrapping with brown paper and muslin cloth; storing in sifted hardwood ashes; dipping in a thick "paint" of fine ashes and water before hanging, and, in cooler weather, leaving them hanging in the smoke house where they are given a fresh shot of smoke about once a week.

Spoilage starts nowhere faster than deep in the bony joints of a ham, far from the salt or brine unless the way is opened with a knife or syringe. I have tried different tricks for overcoming the problem. Commercially they use a special syringe to inject brine deep into the hams.

We have driven a knife in along the bone from each end of a ham and pushed salt deep into the holes. Or you might open those holes with a knife and then inject brine with a basting syringe. But I think the best method for the inexperienced is to open the hams by cutting them to the bone up the inside of the leg. It isn't so pretty, but it is a far sight better than discovering at the dinner table that you goofed in the curing of a fine-looking ham.

Smoking Meat

People seem surprised, even awed, to learn that we smoke our own meats and fish, yet I don't know why. I don't know why smoking, a process as commonplace a hundred years ago as vegetable garden-ing and not nearly so complicated, should come to be considered a high art. Nothing could be simpler than this business of hanging salted meats in smokey places.

There are two types of smoking: *hot* and *cold*. In a hot-smoke the meats are cooked as they hang there absorbing desired colors and flavors. We cold-smoke hams and bacons, but it is only cold in a relative way, since ideally the smoke chamber will be up around 80° or 90°F. (30°C.) when the smoke is on. At colder temperatures absorption of the smoke is slower.

THE SMOKEHOUSE

Our smokehouse is nothing glamorous. It was banged together one afternoon out of used lumber. It looks like any old outhouse from the outside. Inside there are racks near the top from which the meats can be hung. At the bottom sits a cut-off steel drum surrounded by a loose stack of used bricks. The fire goes in the bottom of the drum, right there, three feet below the meats, which is possible because we smoke with sawdust that merely punks along, never producing very much heat.

A smokehouse should be a simple affair. A smudging sawdust "fire" creates lots of smoke but not much heat.

For a draft there is a fist-sized hole punched in the front bottom of the drum, which opens to the front of the smokehouse below the door. A wad of grass regulates the draft. The smokehouse is not absolutely tight. Smoke and moisture leak through numerous cracks.

A smokehouse of this kind should be located in a sheltered spot so that varying winds don't have you running back and forth regulating the draft. Also the amount of smoke filling the house will more likely remain constant if gales of wind can't sing through it.

The meats (and sometimes strips of fatback, too) are hung by strings or stick skewers while they are still dripping. We build a small fire of paper and kindling in one corner of the bottom of the drum and then draw a large pile of sawdust slowly in and around until it has caught. Until the flames have been quelled and the fire reduced to smoldering sawdust, the door of the smokehouse is kept wide open to keep the heat down.

If the sawdust burns too fast you can dampen it down. If it is too wet, dry it out on a barn floor. You will need about four bushels of sawdust for smoking hams or other large chunks of meat.

It must be hardwood sawdust. People slaver over the mention of preferred hardwoods, but I've never noticed a great difference in quality of the finished product. Psychologically applewood is the nuts. Last year we had some seasoned apple branches. We put a large tarpaulin under the sawhorse and cut the branches into inch wheels with the power saw. The combination of wood slices and saw chips made an ideal smoking mixture. Go easy oiling the saw blade at a time like this.

Corncobs are supposed to be very good for smoking, and straw has been mentioned as satisfactory if you've nothing else. Evergreen kindling may be used to start a smokehouse fire but never use a resinous wood or sawdust from pines, spruces, cedars or hackmatacks (larch) for the smoke. They are full of turpentine-like tars and oils.

As mentioned before, the main reason for smoking meats is to give them flavor, so smoking time is a matter of taste. Thinner cuts take the smoke faster. After a full day or two, take a slab of bacon down and fry up a slice taken from the middle. If more smoke is desired, return the severed slabs to the smokehouse.

While the weather remains cool the meats may be left in the smokehouse. Some like to give the meats fresh smoke from time to time.

Meats should not be allowed to freeze and thaw repeatedly. If that is the kind of weather you are having, bring the hams and bacons in and hang them where they will be dry and at about room temperature—this is if they got a full cure. If they only took a mild cure they should go in a deep freeze for storage. *Putting Food By* recommends not attempting to store salted cuts of fat or meat (at 0°F.) for longer than four months. Any longer, they say, and "the salt in the fat causes it to become rancid."

Fully cured meats may be kept at room temperature. If they are being kept through the summer months they should be wrapped in brown paper (large shopping bags will do), be sewn up in cloth bags to keep out the flies, and be hung again in any dry room that does not get terribly hot.

Those who produce the finest fully cured country hams and bacons insist that they are not at their peak of flavor until they have aged a month or two for bacons, and six months to a year for hams.

We never have been able to stay away from them that long.

Bibliography and References

Backyard Livestock. S. Thomas. Countryman Press, 1976.

Behavior of Domestic Animals (The). E.S.E. Hafez, editor. Bailliere, Tindall & Cox, 1962.

Book of the Pig. 3d edition. James Long. "The Bazaar, Exchange and Mart" Office, Windsor House, Bream's Bldgs., E.C. C., 1919.

Butchering, Processing and Preservation of Meat. Frank G. Ashbrook. Van Nostrand Reinhold Co., 1955.

Duke's Physiology of Domestic Animals. 8th edition. M.J. Swenson, editor. Cornell University Press, 1970.

Farm Animal Behavior. A.F. Fraser. Bailliere Tindall, London, 1974.

Feeds and Feeding, Abridged, 16th edition. W.A. Henry and F.B. Morrison. Henry-Morrison Co., 1920.

Feeds and Feeding, Abridged, 9th edition. F.B. Morrison, Morrison Pub. Co., 1961.

"Food and Life." USDA *Yearbook of Agriculture*, 1939.

Genetics of the Pig (The). A.D.B. Smith, O.J. Robinson and D.M. Bryant. University of Edinburgh, 1935.

Handbook of Nature Study. 24th edition, 21st printing. A. B. Comstock, Cornell U. Press, 1970.

History of Domesticated Animals. F.E. Zeuner. Harper & Row, n.d.

Pig (The). William Youatt. Cradock & Co., London, 1847.

Pigs: Breeds and Management. Sanders Spencer. Vinton & Co., London, 1919.

Pigs: Their Breeding, Feeding and Management. V.C. Fishwick. Crosby, Lockwood & Son Ltd., London, 1953.

Restraint of Animals. John R. Leahy and Pat Barrow. Cornell Campus Store Inc., 1953.

Sheep and Pigs of Great Britain. J. Coleman, editor of *The Field* magazine, London, 1877.

Raising Swine. G.P. Deyoe and J.L. Krider. McGraw-Hill, 1952,

Swine Science. M.E. Ensminger. Interstate Publishing Co., 1970.

Symbolic Pig (The). F.C. Sillar and R.M. Meyler. Oliver and Boyd Ltd., Edinburgh and London, 1961.

A Complete Guide to Home Meat Curing. Morton Salt Company, Argo, Illinois.

HEALTH

Swine Diseases. Canada Dept. of Agriculture, Publication #1484. Ottawa, Ontario, 1972.

Livestock Veterinary Health Service (revised edition). Dan W. Shield. The Highsmith Co., Fort Atkinson, Wisc., 1973.

Veterinary Guide for Farmers. G. W. Stamm. Hawthorne Books, Inc., New York, 1963.

Keeping Livestock Healthy. N. Bruce Haynes, Garden Way Publishing, 1978

FEED STANDARDS AND ANALYSIS TABLES

Nutrient Requirements of Swine
Tables of Feed Composition
Both available from: National Research Council, National Academy of Sciences, 2101 Constitution Avenue, Washington, D.C., 20418.

LIVESTOCK SUPPLY HOUSES

C.H. Dana Co., Inc., P.O. Box 300, Hyde Park, Vermont 05655

Ketchum Manufacturing Sales Limited, 396 Berkley Avenue, Ottawa, Canada K2A 2G6

Nasco Agricultural Sciences, 901 Janesville Avenue, Fort Atkinson, Wisconsin 53538

Glossary

Ad lib feeding. Allowing stock free access to unlimited quantities of a feed or feed supplement.

Barrow. Male pig castrated before reaching maturity.

"Blind teat." Small, functionless teat.

Boar. Mature male hog.

Conversion. Ability of a pig to turn food into carcass.

Cryptorchid. Male pig, often sterile because one or both testicles are retained in the abdominal cavity.

"Feeder pig." Pig being reared for pork.

Finish. Last stage of feeding a slaughter pig to meet market requirements in terms of carcass quality. The degree of desired fattening.

Farrow. To give birth. "When is the sow due to farrow?"

Farrow to finish. Type of farm operation that covers all aspects of breeding, farrowing, and raising feeder pigs to slaughter.

Following. Practice of allowing feeder pigs to run behind feedlot cattle so that they may glean unused grains and other nutrients from the cattle manure.

Free choice. See *Ad lib*.

Gilt. Young, sexually mature female prior to birth of her first litter.

Hand-feeding. Restricted feeding of set amounts of food.

Hand-mating. Controlled breeding with confined boars rather than allowing boars to run loose with groups of unbred sows.

Heterosis. The genetic influence of cross-breeding resulting in heightened performance of the offspring.

Hog. Fully grown swine more than a year old.

"Hog down." Practice of allowing pigs to "harvest" a crop in the field.

Hybrid vigor. Heterosis.

In pig. Pregnant sow or gilt.

"Open sow." An unbred sow, not "in pig."

Pig. Young swine, not having reached full growth.

"Piggy sow." Pregnant sow.

Prepotent. Marked tendency of a boar or sow to pass along certain genetically controlled traits.

Ranting. Characteristic behavior of an agitated boar—frothing, champing, nervousness.

Ridgling. See *Cryptorchid.*

Rig. See *Cryptorchid.*

ROP. Record Of Performance—breeding stock tested for feed conversion characteristics.

Runt. Undersized pig. Commonly there is a runt born in every litter.

Self-feeding. Same as *free choice.*

Shoat. Young swine, synonymous with pig. Shoat (also *shote* or *shot*) is an older term.

Sow. Mature female with record of one or more litters.

Stag. When castrated, a mature boar becomes a stag.

Standing heat. Period in sow or gilt's heat during which she will stand still when being mounted.

Swine. Collective term for all age groups within the species.

Appendix

Balancing Your Own Ration

Here is a charted short course through the science of pig feeding. Don't let unfamiliar terms shake the underlying idea that feeding is an art that takes bringing together the right materials and the proper amount of energy to make the pig.

RESTRICTING THE FLOW

Pigs may be starved by an inability to get one or more nutrients, by an inability to digest what they do get (since nutrients don't come in equally digestible forms), or by an imbalance in nutrients (such as occurs when there is insufficient vitamin D for processing minerals). A sick animal may be unable to eat, or if it eats, be unable to use its food.

Lastly there is a restriction of a diet that comes with an excess of bulk. A pig, or any animal for that matter, has a limited gut. It can consume and process only so much quantity of material in a day. Fiber, one of the carbohydrates, is at best only slowly digestible in the pig's system. It is the solid ingedient most responsible for giving feeds bulk. Generally, a feed that "has bulk" takes up more room for a given amount of nutrients than one that is said to "lack bulk."

There are times when it is desirable to slow a pig's growth. Kinder and healthier than simply cutting back its daily ration is to increase the bulk by replacing portions with whole-ground oats, chopped alfalfa, grass or some other high-fiber feed.

COMPARING FEEDS

Just by looking you can make some comparisons and judgments about various feeds' nutritional values. For instance, a pig fed whole oats will have manure that is full of the same oats that came through the system unscathed. Ground oats will not show to the same extent in the manure, and this is not simply a matter of their being less visible. They are more digestible, as are all ground grains, because their tough, outer hulls have been broken in the milling process to expose digestible starches within.

To evaluate and compare feeds scientifically, each is measured in a laboratory for its content of dry matter, which then is broken down into measures of fiber, fat, vitamins, minerals, energy and protein. In some cases the breakdown of proteins is carried further to show amino acid content. Then the feed is given to the pig in a special stall or pen from which wastes can be collected and measured. Original feed value, less the sum of the values of the waste products, equals the value of the food to the pig. The most sophisticated systems allow scientists to collect and measure not only manure, but urine, gasses and body heat as well.

Until recently the energy value of feeds in North America was judged on a basis of Total Digestible Nutrients (TDN) contained. TDN is still used in many books and by feed companies. But measurement in calories is gaining favor (one calorie equaling the energy—in heat—needed to raise one gram of water one degree centigrade).

SOME FEED ANALYSIS TERMS

Here is a brief glossary of terms found in feed analysis tables drawn from laboratory tests.

Dry Matter (DM). The amount of material left in a sample after all the moisture has been driven off.

Air Dry. The weight of a sample that has been allowed to dry to about the 10 percent moisture level.

Ash. The minerals in a sample found by burning away the other ingredients.

Crude Fiber. Most tough carbohydrates in a feed, such as the cellulose. However lignin (which is less digestible than cellulose) is not included in their measure because, unlike the other crude fibers, it is soluble in the test solution used to determine amounts of:

Nitrogen-free Extract (NFE). This is primarily a measure of the more soluble carbohydrates such as the sugars and starches. Sometimes tests are carried further to determine the amount of indigestible lignin in the NFE.

Ether Extract. The fats and oils and related compounds that are soluble in ether.

Protein. Usually listed as a percentage of Crude Protein, the digestible portion of which Morrison says will run about 80 percent. National Research Council analysis tables follow their *protein* with "(N × 6.25)." This is because the protein value of a feed is determined by finding how much nitrogen there is in the sample and multiplying this by 6.25, the reason being that true proteins are 16 percent nitrogen (100 divided by 16 = 6.25).

Actually not all of the nitrogen in a sample is protein or amino acid nitrogen. Some is from compounds called *amids* which a rum infant's infant's gut microorganisms can use to build proteins, but which a pig, with its simple stomach, may only be able to use for carbohydrate-equivalents of energy. Since feeding standards have been written with this discrepency in mind, it isn't critical to know—only interesting.

Energy. Expressed as 1,000 calories (one kilocalorie) per 1,000 grams (Kg).

Gross Energy. All of the energy in a sample as found by burning in a *bomb calorimeter.* The sample is burned by an electrical current within a container inside a tank holding 2,000 grams of water. If the temperature of the water goes up two degrees centigrade it is known that the sample contained 2,000 calories (small *c*) of energy.

Digestible Energy (DE). Caloric value of a sample, less the caloric value of the manure, or Fecal Energy (FE).

Metabolizeable Energy (ME). Caloric value of a sample, less that of the manure, urine and gasses.

Net Energy (NE). Caloric value of a sample less that of the manure, gasses, urine and excess body heat from what is sometimes called the *work of digestion.* (In other words a measure of radiated heat in excess of what the body normally is giving off. This is also sometimes called the *heat increment.*

Total Digestible Nutrients (TDN). As already mentioned, TDN is a measure of the digestible portion of a sample in terms of its energy value. This is carbohydrates, fats (×2.25) and proteins, with the proteins being included because any of this nutrient in excess of what is needed for building can be used as fuel.

Sometimes more than one TDN rating is given a feed to show its value to different kinds of animals. For instance NRC tables give a ground oats sample a TDN rating of 60 percent for swine and 66 percent for cattle, which reflects the ruminant's better use of the fiber hulls.

CALCULATING THE RATION

Through feeding trials with different rations, fairly clear pictures have been drawn showing what nutrients swine need through their various stages of life. The greatest needs are with growing pigs and with sows nursing litters. Then come sows in their last month of gestation and boars in regular and full service.

In professional feeding, the experimentally and practically proven nutritional needs of pigs are listed in terms of quantities of TDN (or Energy), protein, fat, minerals and vitamins required for a balanced daily ration. Foods to fill these requirements are chosen for their nutrients, their price, availability and general suitability to the management and equipment on a particular farm. Then the jockeying begins. Plug in this, scratch that, add a little supplement here, a vitamin there until at last you come up with a formula that fits what the researchers say the pig needs and that you can afford to provide.

One reason why feed companies can provide a balanced ration in a bag for the price (or nearly so) of a single grain, is that all of this information can be programmed into a computer that can almost instantaneously play over the options and come up with a formula that takes the greatest advantage of current market bargains.

This is why it is unlikely anyone will be able to come out ahead by home-mixing grains and supplements and other feeds that have to be bought from a feed store. There is a chance you can do so *if* you grow your own good pasture, grains, peanuts, potatoes or the like. Then all you have to buy outright on the retail market may be a balancing grain or protein and mineral supplement.

No Feeding Standard is going to meet precisely the requirements of one hog. Such standards are only meant to serve as guidelines for successful (economical) hog raising. They are average figures for average animals fed average feeds. Likewise, when using the average figures in Feed Composition Tables to work out a ration, it is a good idea to be liberal and to err on the side of too much rather than too little of any one nutrient.

The feeding standard (page 253) is adapted from National Research Council tables worked out by Canadian and United States researchers. To use the standard, write out the requirements for the animal in question. For example, say that you want to see how to mix up a daily ration for a 75-pound feeder pig.

The standard says the pig will need:

Total Air Dry Feed	Protein	Dig. Energy	Calcium	Phosphorus
4.8 lbs.	15% (.72 lbs.)	7200 Kcal.	.65% (.49 oz.)	.50% (.38 oz.)

If you had a pile of corn, that would be the obvious basis for a ration. What if we fed the pig corn alone? Using the composition table it is found that 4.8 pounds of corn contain:

Feed	Protein	Dig. Energy	Calcium	Phosphorus
4.8 lbs corn	.42 lbs.	7610 Kcal.	.02 oz.	.20 oz.

Feeding Standard

Energy and Protein Requirements of Swine[1]

DAILY REQUIREMENTS

Liveweight range (lb)	Total air-dry feed (lb)	Protein levels (%/diet)	DE (kcal)	TDN (lb)	DE per lb of feed (kcal)	%TDN in ration
Starting, growing and finishing pigs						
10- 25	1.3	22	2100	1.0	1600	80
25- 50	2.8	18	4500	2.2	1600	80
50-110	4.8	15 (16-14)[2]	7200	3.6	1500	75
110-200	7.2	14 (13-12)	10100	5.0	1400	70
Breeding stock						
Bred gilts	4.5	14	6700	3.4	1500	75
Bred sows	4.5	14	6700	3.4	1500	75
Lactating gilts	11.0	15	16500	8.2	1500	75
Lactating sows	12.5	15	18700	9.4	1500	75
Young boars	5.5	14	8200	4.1	1500	75
Adult boars	4.5	14	6700	3.4	1500	75

[1]"Nutrient Requirements of Swine,"Publication 1599, National Academy of Sciences; National Research Council, Washington, D.C. 1968.
[2]Figures in brackets are for rations using corn as the grain with the higher level for the lighter half of this weight range and the lower level for the heavier half.

Reprinted from Agriculture Canada bulletins.

Weights per Bushel and Bushels per Ton of Important Feed Grains

Grain	Lb. per Bushel	Bu. per ton (2,000 lb.)
Barley	48	41.7
Corn, shelled	56	35.9
Corn, in ear	70	28.6
Grain sorghums	55	36.4
Oats	32	62.5
Rye	56	35.9
Wheat	60	33.3

249

Obviously the protein is way off. Calcium and phosphorous also are low. And there is more than meets the eye. Not only is corn low in protein, its quality of protein is down, too. Two essential amino acids, lysine and tryptophan, are seriously lacking. The same deficiencies, in varying degrees, exist in other cereal grains.

Meat tankage is a common protein supplement used to balance swine rations. It may be expensive, but by having around 50 percent of protein it does not take very much to balance a ration.

However, tankage also is lacking tryptophan. It was the realization that tankage alone, or other animal proteins alone, (other than milk) did not make satisfactory supplements to grain rations, that led Wisconsin feed scientists in the early 1900s to come up with *trinity,* or *trio* protein supplement mixes that were combinations of animal and vegetable proteins with alfalfa or another legume added for their compliment of vitamins and minerals. They provided about 40 percent of protein.

At the time the trinity mixes were developed it was not known why they worked so well. Today we know they work because they better provide the needed balance of amino acids, vitamins and minerals. We don't hear so much about them because now there are many ways to mix feeds to meet the same ends. However, the idea of trinity mixtures as supplements to grains is sound and very practical for people with limited experience and resources and who are considering mixing their own pig rations.

TRIO MIXES FOR BALANCED RATIONS

Corn + Meat Scrap or Tankage	= Mix low in tryptophan
Corn + Cottonseed Meal	= Mix low in Lysine & Methionine
Corn + Meat Scrap + Cotton Seed Meal	= Mix adequate in proteins
Corn + Meat Scrap + Cotton Seed Meal + Alfalfa Meal	= Balanced Ration

So, the question to answer in feeding that 75-pound pig is How much corn should be replaced by a protein, mineral and vitamin supplement akin to one of the trio mixtures?

Rather than guess at combinations that may work, most people attack feed balancing problems by way of a fixed goal, such as the desired protein level. In the case of a 75-pound pig it is a 15 percent protein ration.

Corn, we know, has about 7 percent of protein. The trinity mixtures have about 40 percent protein, though to be more precise you may want to figure the percentage of protein in your own concotion of supplements.

USING THE PEARSON SQUARE

The best way for non-mathematicians like myself to find how much of 7 percent and 40 percent protein feeds to mix to form a 15 percent product is to use the Pearson Square.

1. Draw a square and put the desired protein level in the center.

2. Put the protein values of the two ingredients on the left-hand corners.

3. Subtract across the diagonals of the square, the smaller number from the larger in each case, and put the answers on the right-hand corners.

4. Add the right-hand figures.

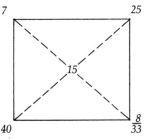

5. Divide this sum into the right hand figures in turn. Round off the answers and multiply by 100 to give the percentage of each for an overall 15 percent mix.

$$\frac{.76 \times 100 \doteq 76\%}{33 \overline{\smash{\big)}\ 25.00}} \quad corn \qquad \frac{.24 \times 100 = 24\%}{33 \overline{\smash{\big)}\ 8.00}} \quad protein\ supplement$$

6. If you are mixing 100 pounds of a 15 percent protein mix you would take 24 pounds of the high-protein supplement and 76 pounds of corn. In each case these answers lie directly across the square from their protein values. These figures (24 and 76) also are the percentages of the ingredients that you would find in any quantity of a 15 pecent protein mixture of the two. And so 4.8 pounds—the pigs daily ration—would have roughly 3.6 pounds of corn and 1.2 pounds of the trio mix.

A trio mix of 50 pounds of tankage, 25 pounds of soybean meal and 25 pounds of ground alfalfa has about 2584 Kcal/Kg. of energy. Divided by 2.2 (number of pounds in a kilogram) gives 1174 Kcal per pound. So:

Feed	Protein % (lbs)	Digestible Energy Kcal/lb	Calcium (oz.)	Phosphorus (oz.)
3.6 Corn	7 .25	5774	.017	.15
1.2 Trio	40 .48	1602	.89	.433
	.73	7376	.907	.583

This ration is well balanced as far as protein and energy are concerned. Close enough. Erred on the side of generosity.

If you were not sure that your pigs were getting their needed calcium and phosphorus it would be possible to include these minerals in the form of bone meal at a rate of about one pound to a hundred pounds of total feed. Also salt should be included at a rate of about a half pound per hundred pounds of total salt feed.

If a trio mix, with its complement of sun-cured alfalfa meal (or other leafy legume or quality hay), is not included in a ration and the pigs are not getting outside on pasture it will be important to see that they get vitamins A and D. These may be provided in the form of a daily teaspoon of cod liver oil for piglets or through the inclusion of a recommended amount of an A & D supplement from a livestock supply store.

Feed Composition Table

FEED	Dry Matter	Dig. Prot.	Dig. Energy (kcal./kg)	Fiber	Fat	Ca	P
Roughages							
Alfalfa hay	87.7	14	2011	29.7	1.6	1.13	.18
Pasture grasses (frt. N. Sts)	22	2.8	643	5.3	1	.12	.07
Grains & grain products							
Barley	89	8.2	3080	5	1.9	.08	.42
Buckwheat	88	8	3026	9	.1	.02	.02
Corn	86.5	7.2	3508	1.8	3.9		
Corn and cob meal	87	5.8	3107	8	3.2	.04	.27
Corn Gluten meal	91	36.8	3451	4	2.3	.16	.4
Cotton seed meal (w hulls)	91.5	34.9	2703	12	2	.16	1.2
Oats	84	9	2629	10	4.8	.08	.3
Peanuts (whole)	92	15.1	4381	21.7	34.4		.3
Rice (grnd., w hulls)	89	5.5	2511	9	1.9	.04	.26
Sorghum	88	8.4	3492	2.1	2.6		
Soybeans	90	31	4048	5	18	.25	.59
Soybean oil meal	89	41.7	3300	6	.9	.32	.67
Grain screenings	91	9.9	2461	16.7	4	.37	.41
Roots							
Beets (Mangels)	10.6	.5	402	.9	.1	.02	.02
Carrots	11.9	.9	430	1.1	.2	.05	.04
Potatoes (fresh)	24.6	1	933	.5	.1		
Potatoes (cooked)	22.5	1.6	863	.7	.1	.01	.05
Sweet potatoes	31.8	.2	1128	1.9	.4	.03	.1
Turnips	9.3	.8	332	1.1	.2	.06	.02
Animal products							
Cottage cheese	21	15.3	889	0	.3	.09	.18
Skimmed milk	9.6	2.7	415	0	.1	.12	.10
Meat meal (tankage, dehyd.)	92	37.1	2475	2	8.1	5.94	3.17
Chicken eggs, whole	34.1	12.8		0	10.6	1.5	
Codfish meal, dehyd.	94	57	2818	1	1.6		
Ocean perch, whole, fresh	19.8	18	800*		.4		
Garden/Orchard by-products							
Apples	17.9	.2	586	1.3	.4	.01	.01
Apple pomace	20.9	.6	630	4.4	1.3	.02	.02
Cabbage, outer leaves	15.8	1.8	467	2.7	.4		
Tomatoes	5.7	.6	216	.6	.4	.01	.03
Garbage							
Hotel/Restaurant garbage	16	2.2	793	.5	4		
Bread, dehyd.	95	9.8	3686	.5	1	.03	.1

*Gross Energy value

Based on figures from National Research Council tables and those from *Feeds and Feeding, Abridged*, F.B. Morrison.

Feed Substitution Table for Swine

Feedstuff	Relative Feeding Value (lb. for lb.) in Comparison with the Designated (underlined) Base Feed Which = 100	Maximum Percentage of Base Feed (or comparable feed or feeds) Which It Can Replace for Best Results	Remarks
GRAIN, BY-PRODUCT FEEDS ROOTS AND TUBERS:[1] (Low and Medium Protein Feeds)			
Corn, No. 2	100	100	Corn is the leading U.S. swine feed, about 50% of the total production being fed to hogs.
			It does not pay to grind corn for growing-finishing pigs, but it should be ground for older hogs.
Barley	90-95	100	Of variable feeding value due to widespread in test wt./bu. Should be ground.
			In Canada, where high quality bacon is produced, barley is considered preferable to corn for finishing hogs.
Beans (Cull)	90	66 2/3	Cook thoroughly; supplement with animal protein.
Carrots (or beets, mangels, or turnips	12-20	25	
Cassava, dried meal	85	33 1/3	
Corn meal	100	20	
Hominy feed	95	50	Hominy feed will produce soft pork if it constitutes more than 1/2 the grain ration.
Millet (Hog Millet)	85-90	50	
Molasses, beet	70-75	20	
Molasses, cane	70-75	20	
Molasses, citrus	70-75	10-20	It takes pigs 5 to 7 days to get used to the bitter taste of citrus molasses.
Oats	70-80	33 1/3-100	For growing-finishing pigs, oats is equal to corn when limited to 1/3 of the ration.
			In Canada, where high quality bacon is produced, oats is sometimes used to finish hogs in order to obtain a lean carcass.
			The feeding value of oats varies according to the test weight per bushel. Grind for swine.
Peanuts	120-125	100	Peanuts are usually fed by hogging-off.
Peas, dried	90-100	100	Normally peas should be fed to swine as a protein supplement. Two tons peas=1 ton grain+1 ton soybean meal.
Potato (Irish)	25-28	25-50	Not palatable in raw state; must be cooked.
Potato (Irish), dehydrated	100	33 1/3	
Potatoes (sweet)	20-25	33 1/3-50	Cooking also improves the feeding value of sweet potatoes.
Potatoes (sweet), dehydrated	90	33 1/3	
Rice (rough rice)	80-85	50	Rice should be ground.
Rice bran	100	33 1/3	If more than 1/3 of the grain consists of rice bran, soft pork will result.
Rice polishings	100-120	33 1/3	Limited because feed becomes rancid in storage and soft pork will be produced.
Rice screenings	95	50	
Rye	90	50	Should be limited because it is unpalatable. Grind for swine.
Sorghum, grain	90	100	All varieties have about the same feeding value. Grind when hand-fed.
Sunflower seed	100	50	
Wheat	105	100	Feed whole if self-fed; grind if hand-fed.
Wheat bran	75	15-25	Bran is particularly valuable at farrowing time.
			In Canada, where high quality bacon is produced, 15 to 25% wheat bran is sometimes incorporated in the finishing ration in order to obtain a lean carcass.
Wheat flour middlings	103	20	
Wheat standard middlings	85-100	25-50	Combine with animal protein and limit to 1 lb. per head daily.
Wheat red dog and wheat white shorts	115-120	25	

Feedstuff	Relative Feeding Value (lb. for lb.) in Comparison with the Designated (underlined) Base Feed Which = 100	Maximum Percentage of Base Feed (or comparable feed or feeds) Which It Can Replace for Best Results	Remarks
PROTEIN SUPPLEMENTS			
			In general, animal proteins should comprise part of the total protein supplement of swine; especially in confinement and for young pigs and gestating-lactating sows; they may comprise more it they are a cheaper protein source than plant proteins. Plant proteins may comprise as much as 90% of the protein supplement, provided the ration is adequately fortified with vitamins and minerals.
Tankage (60%)	100	100	
Buttermilk, dry	90-105	100	
Buttermilk, liquid	15	100	Pound for pound, worth 1/10 as much as dried buttermilk.
Buttermilk, semisolid	33 1/3-50	100	Pound for pound, worth 1/3 as much as dried buttermilk.
Copra meal (coconut meal)	50	25	
Corn gluten meal (gluten meal)	50-75	50	
Cottonseed meal (41%)	50-75	33 1/3	Except where the new screw-processed cottonseed meal is used, high level may produce gossypol poisoning and the level of cottonseed meal in swine rations should never exceed 8 to 9% of the total ration.
Fish meal (63%)	105-110	100	
Linseed meal (35%)	50-75	25-50	
Malt sprouts	100	10	Malt sprouts contain a growth factor(s). They result in increased feed intake and gain.
Meat and bone scraps (45-50%)	95-100	100	
Meat scraps (50-55%)	100	100	
Peanut meal (41%)	75-80	50	Becomes rancid when stored too long. High levels may produce soft pork.
Peanuts	60-70	50	Peanuts are usually fed by hogging-off.
Peas, dried	50	50	
Shrimp meal	90-100	50	
Skim milk, dried	90-120	100	In limited amounts, more valuable than tankage for young pigs.
Skimmed milk, liquid	15	100	Pound for pound, worth 1/10 as much as dried skim milk.
Soybean meal (41%)	75-85	50	Soybean meal is of better quality than the other protein-rich plant protein supplements.
Soybeans	70-75	50	
Whey, dried	45	100	
Whey, liquid	30	50	Worth 1/2 as much as skim milk.
PASTURES AND DRY LEGUMES			
Pasture, good		5-20% of grain, and 20-50% of protein supplement.	Pasture and dry legumes are sources of good quality proteins, of minerals, and of vitamins. Thus, swine should have access to either pasture or ground legume.
Alfalfa meal		It can replace all of pasture, in dry-lot rations.	For confinement rations, include 5-10% alfalfa in ration of growing-finishing pigs, and 15-35% in ration of gestating-lactating sows. In Canada, where high quality bacon is produced, up to 25% alfalfa meal is sometimes included in the finishing ration in order to obtain a lean carcass.

[1]Roots and tubers are of lower value than the grain and by-product feeds due to their higher moisture content.

Reprinted from Ensminger, *Swine Science*, 4th edition, 1970. Published by The Interstate Printers and Publishers, Inc., Danville, Ill.

Suggested Feed Combinations for Various Classes of Hogs

Class of hogs	Corn (Lbs.)	Barley (Lbs.)	Wheat (Lbs.)	Oats (Lbs.)	Rye (Lbs.)	Grain sorghum (Lbs.)	Tankage or fish meal[1] (Lbs.)	Cotton seed meal[2] (Lbs.)	Linseed meal[2] (Lbs.)	Alfalfa meal (Lbs.)	Soybeans[2] (Lbs.)	Soybean meal (Lbs.)
Growing—weaning to 100 pounds (dry lot)	80	—	—	—	—	—	10	—	5	5	—	—
	—	85	—	—	—	—	5	5	—	5	—	—
	—	—	90	—	—	—	4	2	2	2	—	—
	65	—	—	20	—	—	5	—	—	5	—	5
	80	—	—	—	—	—	7	—	—	5	8	—
	—	—	—	—	—	85	5	5	—	5	—	—
Growing—weaning to 100 pounds (on pasture)	—	40	25	25	—	—	5	5	—	—	—	—
	55	—	—	—	30	—	5	—	—	5	—	5
	—	90	—	—	—	—	5	—	5	—	—	—
	85	—	—	—	—	—	9	—	6	—	—	—
	30	40	20	—	—	—	5	—	—	—	—	5
	—	92	—	—	—	—	4	—	4	—	—	—
Fattening hogs over 100 pounds (dry lot)	—	45	25	25	—	—	2	3	—	—	—	—
	45	—	—	—	45	—	5	—	5	—	—	—
	89	—	—	—	—	—	5	—	—	—	6	—
	82	—	—	—	—	—	4	—	4	10	—	—
	21	50	20	—	—	—	5	—	—	4	—	—
	—	—	—	30	—	62	4	—	—	—	—	4
Fattening hogs over 100 pounds (on pasture)	47	—	—	—	47	—	6	—	—	—	—	—
	—	—	—	—	—	92	4	—	4	—	—	—
	90	—	—	—	—	—	5	—	—	—	—	5
	70	—	—	25	—	—	5	—	—	—	—	—
	—	96	—	—	—	—	4	—	—	—	—	—
	20	50	25	—	—	—	2	—	3	—	—	—
	87	—	—	—	—	—	3	—	—	—	10	—
Brood sows and boars	88	—	—	—	—	—	6	—	3	3	—	—
	65	20	—	—	—	—	5	—	—	10	—	—
	65	—	—	30	—	—	5	—	—	—	—	—
	—	—	—	—	—	82	4	4	—	10	—	—
	—	—	90	—	—	—	5	—	—	5	—	—
	—	62	—	30	—	—	3	—	—	—	—	5
	85	—	—	—	—	—	2	—	—	5	8	—

[1]Shrimp meal may be substituted for tankage or fish meal in the above mixtures. Skim milk or buttermilk may be substituted for tankage or fish meal. Approximately 11 pounds of skim milk or buttermilk will replace 1 pound of tankage.
[2]Linseed meal, soybeans, or cottonseed meal may be used interchangeably.

Reprinted from *USDA Yearbook of Agriculture, 1939.*

Suggested Formulas for Protein Mixtures[1]

Formula No.	Tankage (lbs.)	Linseed meal (lbs.)	Alfalfa meal (lbs.)	Cotton-seed meal (lbs.)	Soybean meal (lbs.)	Ground soybeans (lbs.)
1	50	25	25	—	—	—
2	50	—	25	25	—	—
3	25	—	25	—	—	50
4	25	—	25	50	—	—

(Continued next page)

Suggested Formulas for
Protein Mixtures, (Cont'd)

Formula No.	Tankage (lbs.)	Linseed meal (lbs.)	Alfalfa meal (lbs.)	Cotton- seed meal (lbs.)	Soybean meal (lbs.)	Ground soybeans (lbs.)
5	25	—	25	—	50	—
6	30	—	35	35	—	—
7	25	—	50	—	25	—

[1]Either fish meal or shrimp meal may be substituted for tankage in the above mixtures. Skim milk or buttermilk may be used instead of tankage. When used as the only protein supplement, 2 pounds of milk may be used to 1 pound of grain.

Reprinted from *USDA Yearbook of Agriculture, 1939.*

Conversion Factors for
Metric System

Imperial units	Approximate conversion factor	Results in:
Linear		
inch	×25	millimeter (mm)
foot	×30	centimeter (cm)
yard	×0.9	meter (m)
mile	×1.6	kilometer (km)
Area		
square inch	×6.5	square centimeter (cm^2)
square foot	×0.09	square meter (m^2)
acre	×0.40	hectare (ha)
Volume		
cubic inch	×16	cubic centimeter (cm^3)
cubic foot	×28	cubic decimeter(dm^3)
cubic yard	×0.8	cubic meter (m^3)
fluid ounce	×28	milliliter (ml)
pint	×0.57	liter (*l*)
quart	×1.1	liter (*l*)
gallon	×4.5	liter (*l*)
bushel	×0.36	hectoliter (hl)

(Continued next page)

Conversion Factors for
Metric System, (Cont'd)

Imperial units	Approximate conversion factor	Results in:
Weight		
ounce	×28	gram (g)
pound	×0.45	kilogram (kg)
short ton (2000 lb)	×0.9	tonne (t)
Temperature		
degrees Fahrenheit	(°F-32)×0.56 or (°F-32)×5/9	degrees Celsius (°C)
Pressure		
pounds per square inch	× 6.9	kilopascal (kPa)
Power		
horsepower	×746	watt (W)
	×0.75	kilowatt (kW)
Speed		
feet per second	×0.30	meters per second (m/s)
miles per hour	×1.6	kilometers per hour (km/h)
Agriculture		
gallons per acre	×11.23	liters per hectare (l/ha)
quarts per acre	×2.8	liters per hectare (l/ha)
pints per acre	×1.4	liters per hectare (l/ha)
fluid ounces per acre	×70	milliliters per hectare (ml/ha)
tons per acre	×2.24	tonnes per hectare (t/ha)
pounds per acre	×1.12	kilograms per hectare (kg/ha)
ounces per acre	×70	grams per hectare (g/ha)
plants per acre	×2.47	plants per hectare (plants/ha)

Index

Abcess, 178
Air temperature, effect on feed
   conversion (*table*), 73
Air temperature, effect on rate of gain
   (*table*), 73
Amino acids, 96–98
Anemia (*see* Piglets, Iron shots)
Antibiotics (*see* Pre-starters and
   antibiotics)
Artificial insemination, 136
atresia ani, 44
Atrophic rhinitis, 178–79

Bacon cures, 235–37
Balling gun, 173
Bedding, 83; 147–48
Behavior
   biting, 34
   cleanliness, 40–41
   Eating, 88
   eyesight, 35
   hearing, 36
   mobility, 37–38
   scratching, 41–42
   social, 38–39
   voice, 36–37
   wallowing, 39–41
   weather forcasting, 41
Biting, 34
Boar, wild, 7–11
   domestication, 13–15
   history, 7–8

   relationship to pig, 9–12
   reproduction, 10–12
   temperament, 12
Breed associations, 52
Breed differences, 30–31; illus. 25
Breeding, 50–52; (*see also* Farrowing)
   boar, 52
   culling, 148–49
   gestation, 137
   hand-mating, 134–37
   heat detection (*graph*), 132–34
   heat periods, 131–32
   records, 149
   reproduction failures, 192–93
   selecting stock, 50–51
Breeding crate, 136
Bristles, 18
Brucellosis, 179, 182
Bulk, 244
Butchering; *see also* Slaughtering
   equipment, 196–200
   preparation, 196–201
Buying a pig
   barrows, 49
   health, 47–49
   how many, 46
   price, 49
   when, 45
   where, 46–47

Calcium and phosphorus (*table*), 99–100
Calling, 55–56

Carbohydrates, *see* Fats and
  carbohydrates
Carcass
  percentages of parts according to live
    weight (*table*), 85
  quality, 89–91
  soft and hard pork, 90–91
Cash investment, 2–3
Castration, 156–60
Chilling, 218–19
Choking, 192
Cholera, 182
Clearing land, 19; 119–20
Coat, 42–43
Colibacillosis, 179
Colostrum, 144–45
Conformation (*illus.*), 42
Copper deficiency, *see* Iron deficiency
Creeps, 82–83
Crytorchidism, 44
Culling sows and boars, 148–49
Cures, ham and bacon, 235–37
Curing, 232–37
  wet and dry cures, 233–35

Dancing pig disease, 185–86
Deficiency diseases, 187–89
Devil-pig, 27–29
Diagnostic chart, 180–81
Digestion, 87–89
Diseases, 178–191
Diamond skin disease, *see*
  Swine erysipelas
Docking, 155–56
Drenches, 173

Energy and protein requirements
  (*table*), 249
Equipment
  bedding, 83; 147–48
  creeps, 82–83
  troughs, 79–82
  watering systems, 81–82
Escaped pigs, 59
Eyesight, pig, 35

Farrowing, 137–48
  bedding, 147–48

care of piglets, 142–43
milk formula for orphans, 145
nursing, 143–44
pens and crates, 146
weaning, 145–46
Fat probing, 162–63
Fats and carbohydrates, 95–96
Feed
  comparative value (*table*),
    120–21
  comparing scientifically, 245
  conversion efficiency according to
    weight (*table*), 106
  fish and fish by-products, 127
  following cattle, 127
  formulas for protein mixtures (*table*),
    256
  garbage, 122–24
  grower ration, 104–6
  home-grown crops, 126
  home-mixed rations, 113
  inexpensive grains and mill wastes,
    120–21
  mast, 126
  milk and milk products, 129–30
  pasture, 116–20
  pre-mixed rations, 103
  pre-starters and antibiotics, 103–4
  roots, 124–26
  starter ration, 104
  vitamins and minerals, 130
Feed analysis terms, 245–47
Feed and water consumption (*table*), 95
Feed combinations (*table*), 256
Feed composition (*table*), 253
Feed conversion, 85–86; *see also* Air
  temperature, effect on feed conversion
Feed substitution (*table*), 254–55
Feeding
  ad lib *vs.* controlled, 108
  boars, 112
  breeding stock, 109–13
  farrowing pigs, 110–11
  hand feeding, 79, 108
  home-mixed rations, 113
  piglets, 112
  troughs, 79–82
  wet vs. dry, 107–8

Feeding, hand, 79, 108
Fencing, 66, 75–76
Fever, 192; *see also* Temperature, taking; Diagnostic chart
Fighting, 165–67
Foot trimming, 163
Freaks, 43–44

Garbage, 122–24
  composition (*table*), 122
  cooking, 123–24
  grading system, 123
Gestation, 137; (*table*), 138
Gutting, 209–212
Grains and mill waste, 120–21
Grower ration, 104–6

Ham cures, 235–37
Hand-feeding, 79; 108–9
Head cheese and scrapple, 230–32
Hearing, 36
Heat detection, 132
Heat periods, 131
Heat stroke, 72–73; 191
Heritability estimates (*table*), 26
Hernia, 44
Heterosis, 52
Hog, parts of (*illus.*), 43
Hog holder (*illus.*), 60–61
Home-mixed rations, 113
Housing, indoor
  doors, 68
  fences, 66
  floors, 66–68
  lighting, 69
  pen, 66
  size, 63–66
  total confinement, 69–71
  ventilation, 71
  walls and partitions, 68
Housing, outdoor
  fencing, 75–76
  hutch, 71–72
  mature hogs, 78
  mixing animals, 76–77
  tethering, 77–78
  wallows, 77
  yards and pastures, 74–75

Housing requirements (*table*), 65
Hurdles, 167–68
Hutch, 71–72
Hybrids, 52
Hypoglycemia, 185–86

Infectious diseases, 178–87
Injections, 173–75
Injuries, 191–92
Intelligence, 31–32
Iodine deficiency, 187
Iron deficiency, 99; 187–88
Isolation, 172

Killing, 201–204

Lameness, 191
Lard production, history, 17–18
Lard, rendering, 227–29
Lead poisoning, 190
Leptospirosis, 182
Lice, 186

Mange, 186
Manure, pig (*table*), 4–5
Manuring area, 168
Masting, 126
Meat cutting (*photos*), 223–27
Meat processing equipment, 22
Medication
  injections, 173–75
  oral, 173
  topical antiseptics, 175
Medication, administering 173–76
Mercury poisoining, 190
Metric conversion (*table*), 257–58
Milk and milk products, 129–30
Milk formula, 145
Milk production (*graph*), 110
Mill wastes, *see* Grains and mill wastes
Mineral supplements (*tables*), 130
Mixing animals, 76–77
MMA, 182–83
Mold poisoning, 190

Needle teeth, 88; 150–52
Noose, 60

Nursing, 143–44; *see also* Milk production
Nutritional imbalance, 187–89
Nutritional needs, 93–101
  amino acids, 96–98
  calcium and phosphorus, 99–100
  fats and carbohydrates, 95–96
  protein, *see* Amino acids
  vitamins and minerals, 98–101
  water, 94–95

Packaging meat, 227
Parasites, 171–72; 186–87
  internal, 171–72
Pasture, 116–20
Pasture crops (*table*), 117
Pasturing dangers, 120
Pearson square, 251
Peccary, 20–21
Phosphorus, *see* Calcium and phosphorus
Pig
  behavior, *see* Behavior
  body type (*illus.*), 26
  intelligence, 31–32
Pig, for sport, 15–16
Pig, history, 20; 27–29; *see also* Boar
Pig, raising,
  economics, 5–6
  expenses (*table*), 4
  labor, 3–4
  manure, 4–5
  reasons, 1–6
  short-term commitment, 1–2
  transporting, 53–59
Pig, uses other than meat, 15–21
Piglets
  care of newborn, 142–43
  castration (*illus.*), 156–59
  clipping needle teeth, 150–52
  docking, 155–56
  ear notching, 153–56
  iron shots, 152–53
  marking, 150
  nursing, 143–44
  tattooing, 155
  transporting, 53
  weighing, 160–61
Pigskin, 18

Pitch poisoning, 191
Pneumonia, 183
Poisoning, 189–91
Pork, cookbook, 232
Pork, religious laws against eating, 14–15
Pork cuts (*illus.*), 221
Pre-mixed rations, 103–4
Pre-starters and antibiotics, 103–4
Protein, *see* Amino acids
Protein levels (*table*), 105
Protein mixtures, formulas, 256–57

Ration, calculating 247–52; *see also* Feed
Records, 149
Reproduction, *see* Breeding
Reproduction failures, 192–93
Restraint, 60–62
Rickets, *see* Vitamin D deficiency
Ringing, 164–65
Rodent (anti-coagulant) poisoning, 190
Rooting, 33–34; 118–19
Roundworms, 187

Salt poisoning, 191
Salt pork, 228
Sausage, 229–30
Scalding, 200–201; 205–8
Scours, baby pig, 179
Scraping, 205–8
Scrapple, 231; *see also* Head cheese
Selenium, *see* Vitamin E deficiency
Skeleton, 32–33
Skinning, 205
Slaughtering, 202–219; *see also* Butchering
SMEDI, 183
Smokehouse, 237–39
Smoking meat, 237–39
Snakes and pigs, 18
Soft pork, 90–91
Sprinklers, 77
Standing heat, 132
Starter ration, 104–6
Sticking, 203–4
Stomach worms, 186
Sunburn, 191–92
Swine breed associations, 52
Swine (hemorrhagic) dysentery, 184
Swine erysipelas, 184

Swine flu, 184–85
Swine management, 21–27
    breeding goals (*illus.*), 24–27
    early breeding, 23–24
Swine pox, 185
Swineherding, 21–23
Symptoms of illness, 176; *see also* Diagnostic chart

Tattooing, 155
Temperature, taking, 170–71
Temperature, subnormal, 192; *see also*
    Temperature, taking
Tethering, 77–78
Total confinement, 69–71
Transmissable gastroenteritis, 185
Transporting piglets, 53
Transporting pigs, 54–55
    driving, 56–59
    herding, 56
Trichinosis, 187
Troughs, 79–82
    materials, 80–81
Tusking, 168

Vesicular exanthema, 185
Vitamin A deficiency, 188–89
Vitamins and minerals, 98–101
Vitamin B deficiency, 189
Vitamin D deficiency, 188
Vitamin E deficiency, 189
Voice, 36–37

Wallows, 77
Water, 94–95
Watering systems, 81–82
Weaning, 145–46
Weighing pigs, 161
Weight gain, 84–89; *table,* 106
    digestion, 87–89
    feed conversion, 85–86
    intestines, 86–87
Weights per bushel of grain (*table*), 253
Wet vs. dry feed, 107–8
Worming, 171–72
Worms, 187

Yellow fat disease, 128

Zinc deficiency, 188

Other Storey Titles You Will Enjoy

The Family Cow, by Dirk Van Loon. Practical, fully illustrated chapters with accurate information on buying, behavior, nutrition, breeds, handling, milking, calving, and growing feed crops. 272 Pages. Paperback. ISBN 0-88266-066-7.

A Guide to Raising Chickens, by Gail Damerow. An informative book that enables both beginning and experienced chicken owners to be successful raising chickens. 352 Pages. ISBN 0-88266-897-8.

A Guide to Raising Llamas, by Gale Birutta. A comprehensive handbook covering behavior, training, facilities, showing, health care, breeding, and birthing. 304 Pages. Paperback. ISBN 0-88266-954-0.

Keeping Livestock Healthy: A Veterinary Guide to Horses, Cattle, Pigs, Goats & Sheep, by N. Bruce Haynes, D.V.M. Provides in-depth tips on how to prevent disease through good nutrition, proper housing, and appropriate care. Includes an overview of the dozens of diseases and technologies livestock owners need to know. 352 Pages. Paperback. ISBN 0-88266-884-6.

A Guide to Raising Pigs, by Kelly Klober. Offers small-scale farmers clear, illustrated information about every aspect of pig raising, including raising pigs as a business. 320 Pages. Paperback. ISBN 1-58017-011-0.

These books and other Storey books are available at your bookstore, farm store, garden center, or directly from Storey Publishing, Schoolhouse Road, Pownal, Vermont 05261, or by calling 1-800-441-5700. www.storey.com